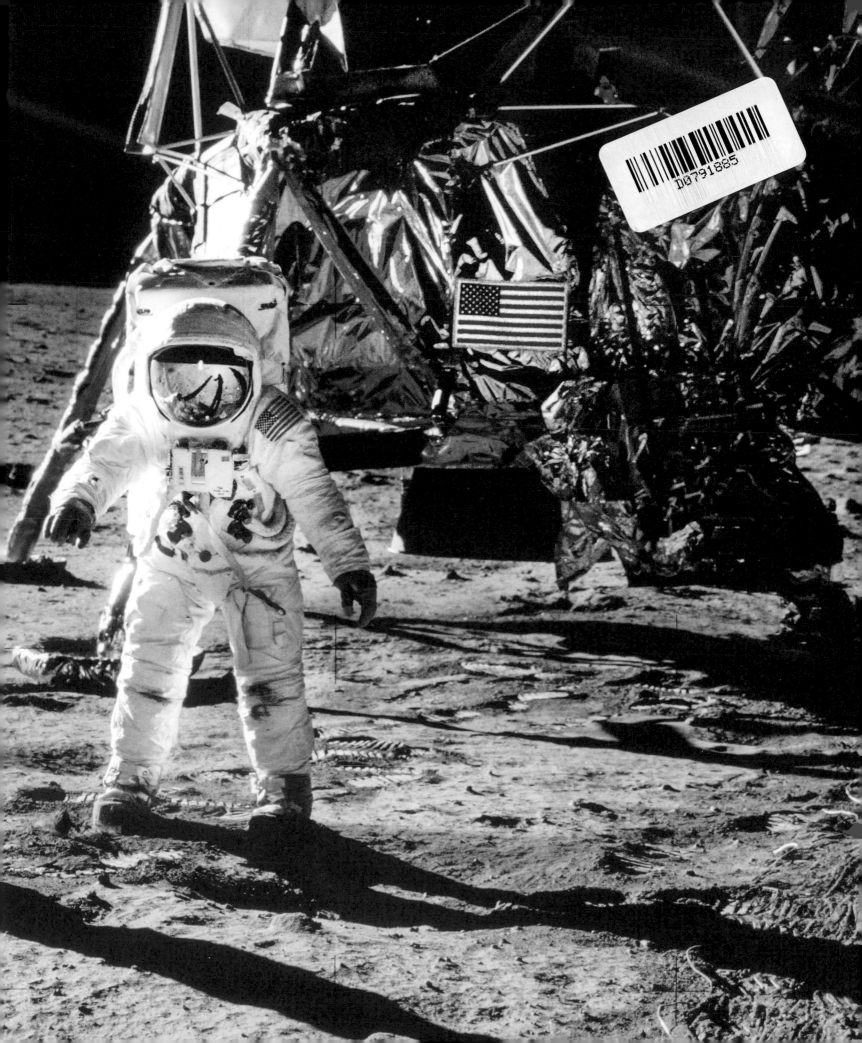

FIRST ON THE
MOON

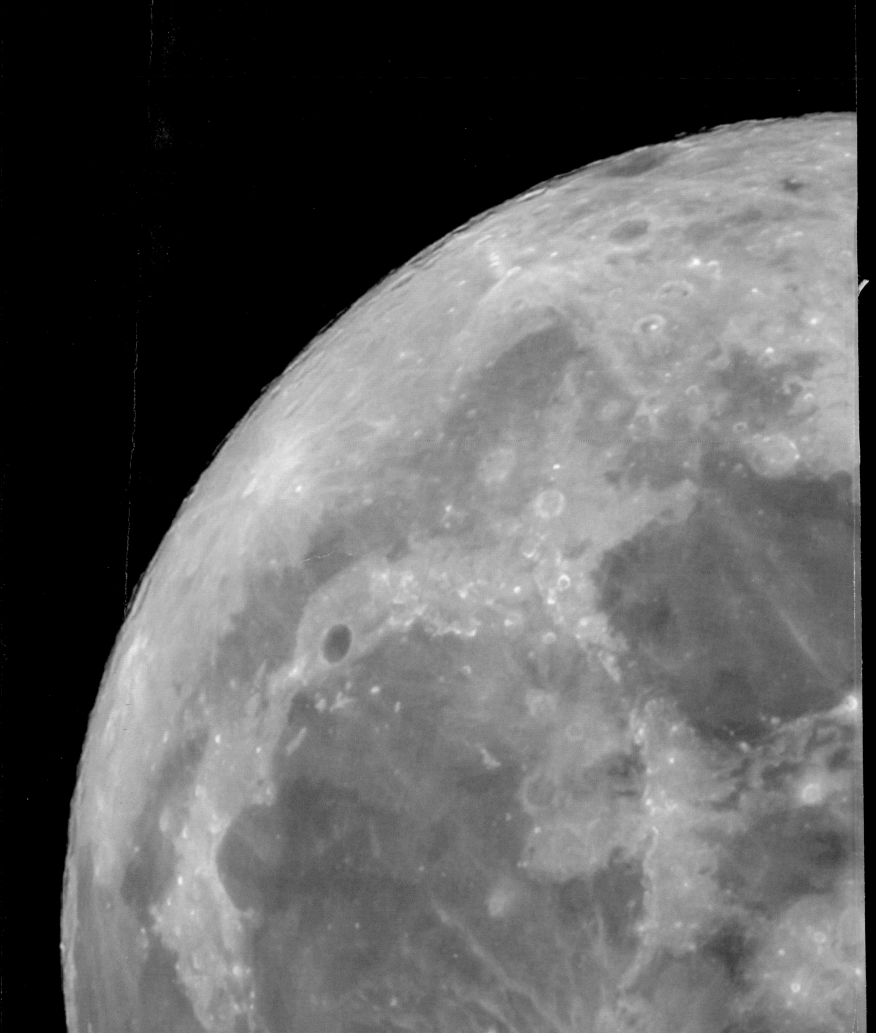

FIRST ON THE
MOON

THE APOLLO 11 50TH ANNIVERSARY EXPERIENCE

ROD PYLE

FOREWORD BY
BUZZ ALDRIN

STERLING
New York

Distributed in Canada by Sterling Publishing Co., Inc.
c/o Canadian Manda Group, 664 Annette Street
Toronto, Ontario M6S 2C8, Canada
Distributed in the United Kingdom by GMC Distribution Services
Castle Place, 166 High Street, Lewes, East Sussex BN7 1XU, England
Distributed in Australia by NewSouth Books
45 Beach Street, Coogee, NSW 2034, Australia

For information about custom editions, special sales, and premium and corporate purchases,
please contact Sterling Special Sales at 800-805-5489 or specialsales@sterlingpublishing.com.

Manufactured in China

sterlingpublishing.com

Interior design by Ashley Prine, Tandem Books
Image Credits — see page 188

Dedicated to the nearly half-million men and women who made our first expeditions to the Moon possible. You performed miracles in less than a decade.

CONTENTS

Foreword ix

Introduction x

CHAPTER 1
"PROGRAM ALARM!"
1

CHAPTER 2
RACE TO THE MOON
9

CHAPTER 3
MEN FOR THE MOON
19

CHAPTER 4
HOW DO WE DO THIS?
27

CHAPTER 5
THE REAL STUFF
39

— APOLLO'S CHILDREN —
58

CHAPTER 6
MOON MACHINES I:
THE SATURN V
63

CHAPTER 7
MOON MACHINES II:
THE LUNAR MODULE
79

CHAPTER 8
T-MINUS ZERO
91

CHAPTER 9
ARRIVAL
107

CHAPTER 10
DOWN AND SAFE
117

CHAPTER 11
MOONWALKING
129

CHAPTER 12
COMING HOME
163

CHAPTER 13
AFTERMATH
171

EPILOGUE
BACK TO THE MOON
177

Acknowledgments 185

Notes 186

Image Credits 188

Index 189

FOREWORD

On July 20, 1969, at 109:43:10 hours' mission elapsed time, I stepped onto the lunar surface. The Apollo 11 mission that delivered us to that forbidding, rocky world was the culmination of a decade of intensive development and stunning human creativity, and that first landing forever changed how humanity looks at both itself and the heavens.

Neil Armstrong and I spent over two hours on the surface of the Moon. We set up a number of scientific instruments, spoke to the president of the United States, collected soil and rock samples, and conducted experiments in mobility, the physical nature of the lunar surface, and much more. We learned more about the Moon during our short visit than had been gleaned in all of human history. Five more Apollo crews explored the Moon, building on the knowledge gained from our first brief visit.

Now we stand on the threshold of a new space age. After decades working in Earth orbit with the space shuttle and the International Space Station, humanity is seriously considering how to send humans to explore what lies beyond. New rockets from companies like SpaceX, Blue Origin, and United Launch Alliance are poised to open new vistas in the heavens. Other nations and several private companies are pursuing space exploration with their own programs, and increasingly in partnership. My own designs for cycling spacecraft to efficiently take humans to Mars are gaining acceptance. The future of human space exploration and development is brighter than ever.

But this all began with those early, tentative forays into the heavens during the space advances of the 1960s, which led to those first steps on the moon in 1969. I'm proud to have been a part of that incredible journey, as are my fellow astronauts who followed us to the Moon. I'm also grateful to the 400,000 men and women who built the Apollo program, the American public who supported it, and the amazing accomplishments in space that have followed. Now it's time to step up our commitment to space, and I'm pleased to say we are doing so.

As I think about what space adventures await us, it's hard to believe that it has been five decades since Neil and I walked on the Moon. I may be biased, but I think it was the high point of the twentieth century with regard to human achievement, and it shows what we can accomplish when we work together toward a great and worthy goal. My memories of that first journey and landing are still clear and will be with me forever. I hope you enjoy reliving those wonderful moments with me in this book.

Ad Astra!

Buzz Aldrin
Astronaut and Global Space Statesman
National Space Society Board of Governors

OPPOSITE: Edwin E. "Buzz" Aldrin suited up and doing a final check of his communication system before boarding for the Apollo 11 mission.

INTRODUCTION

NASA and I were born just two years apart. I entered this world in 1956, and America's space agency was founded in 1958. My own lifelong fascination with space exploration and science grew in parallel with the Gemini program of the mid-1960s, and as it evolved into the Apollo flights, I became hooked. Mankind was going to the Moon, and I read everything I could get my hands on about the program.

Even to my young mind, this was humanity's greatest undertaking: the dramatic extension of our species beyond Earth and the opening of the final frontier. The Apollo program was the first step in the exploration of the cosmos I'd seen just a few years earlier on science-fiction shows like *Star Trek* and *Lost in Space* and read about from writers such as Robert Heinlein and Ray Bradbury. And, in contrast to many of those tales, which represented a kind of "strap it on and go" approach to spaceflight, NASA executed its programs with calm efficiency, discipline, and determination. And, of course, America had to beat the mighty

President Dwight D. Eisenhower (center) signed NASA into existence in 1958. Here, he is ceremoniously appointing T. Keith Glennan (right) and Hugh L. Dryden (left) its administrator and first deputy administrator, respectively, at the agency's opening.

Soviet Union, who for so had long bested the US in spaceflight firsts, in their decade-long competition to land people on the Moon.

But for a young person desperately interested in spaceflight, the 1960s were a trying decade. At the time, there were just three dominant television networks in the United States and only a handful of periodicals that routinely covered spaceflight and NASA. The books available on the topic were few, and those you could find were aimed at either young children or adult readers—there was not yet the plethora of reading material that would emerge during the following decade. During the long and challenging time leading up to that first Moon landing, just a few

A broadcast image from the Apollo 11 Moon landing with a somewhat surreal caption.

popular magazines and the occasional TV news brief covered the events in a manner meant for kids my age. It was hard to get my hands on a lot of material I could really understand, but this changed during the flights of Apollo 8 and Apollo 11, as the media swept in to cover the story of the century. Once we were on our way to the Moon, instead of just planning, the world became glued to TV sets and radios, following these missions of drama and discovery moment by moment.

The Apollo 11 mission is undoubtedly one of the most famous and renowned accomplishments in human history. I was twelve years old when a Saturn V rocket launched the Command Module *Columbia* and the Lunar Module *Eagle* on a path to the Moon. I dragged the two televisions in our house into the living room (no small task in an era when TVs were the size of dishwashers). Both were black and white, befitting the gray and grainy transmissions from the mission. With my dual-screen setup I was able to view two networks at once—my own private Mission Control.

The evening of that first short Moonwalk came and went all too quickly. I'd followed every minute of the mission, from launch preparation to the dramatic narration of liftoff by Jack King, NASA's launch

commentator, to Walter Cronkite's rapt updates of the mission on CBS. As Cronkite narrated the dangerous first landing on the Moon, assisted by Wally Schirra, a recently retired astronaut of much renown, I choked up along with them as the first transmission came back from the lunar surface—"Houston, Tranquility Base here. The *Eagle* has landed." Across the bottom of the screen scrolled a title, for the first time ever, "Live from the surface of the Moon." Those words alone were enough to strain the imagination. Schirra said, "Oh, Jesus." Cronkite huffed a breath and said, "Wally, say something. . . . I'm speechless!"

It was glorious.

What few in the public really understood at the time was just how fragile the space hardware was, and how dangerous the lunar missions were. The Apollo program operated right at the edge of 1960s technology, and the incredible string of successes—even the safe return of the crew of Apollo 13—is a testament to the unwavering dedication of a young and vibrant NASA. The fact that we have not returned humans to the Moon since 1972 is one indication of the complexity and risk associated with the task. We owe those pioneers of the space age a lot.

This intense interest and media coverage of NASA's missions unfortunately would not last. By the middle of the Apollo 14 Moonwalk, when it was clear that there would be no Apollo 13–style life-and-death spectacle, the TV networks switched over to "regularly scheduled programming." Imagine that! While men were exploring the Moon, soap operas and sitcoms dominated the airwaves. CBS, NBC, and ABC switched over to *As the World Turns*, *Days of Our Lives*, and *General Hospital*. The smaller, local networks ran *I Love Lucy* reruns.

I've never really forgiven them.

Despite the dizzying and dazzling heights we had achieved, by 1972, it was all over. Apollos 18, 19, and 20 had been cancelled, allegedly for fiscal reasons, and the Nixon administration had already tossed the Apollo hardware into museums and onto the scrapheap of history. One more Saturn V would launch Skylab, followed by three Saturn 1Bs carrying crews to the station, and a Saturn 1B and Command Module would link up in orbit with a Soviet spacecraft, but the monumental missions to the Moon were over. None of us could have imagined it would come to such an abrupt halt when the massive program was building up, but that's what happened. The brilliant engineering and overwhelming technological prowess that fueled the Apollo program were channeled into the space shuttle, and for the next four decades, NASA explored Earth orbit with steadily declining budgets.

Thankfully, though, popular interest in space is on the upswing, and at the fiftieth anniversary of the first Moon landing, we again stand on the brink of a new space age. A number of nations have plans to put humans into space within the next few years: China will build a large modular space station, India will soon fly its first crew into orbit, and NASA and its international partners will begin the assembly of a lunar orbiting station. Not to mention the private entrepreneurs, the brilliant and determined billionaires of the new space age who will soon fly crews into orbit, to the Moon, and, if Elon Musk has his way, to Mars.

Unlike during my childhood, today we have thousands of specialty media outlets serving space and science audiences, from online venues to specialty cable networks to niche print and e-book publishers. The new space age we are entering will be covered as never before, and it's a blessing for those of us who love space exploration. We are at the cusp of a bright, new era, and I'm thrilled to see it arrive at last.

But it all really started with the voyages of Apollo, the first time humanity left the cradle of Earth orbit. The bravery of the men who traveled to the Moon, and the dedication of the half-million souls who labored to send them there, will always live on in my heart. Some may remember the Kennedy administration as "Camelot," and for me, that shining time when humanity first reached beyond our planet to another world is the twentieth century's brightest moment.

It is my honor and my privilege to share the magnificent voyage of Apollo 11 with you in this fiftieth-anniversary celebration of the mission. From its bold inception, through its seemingly insurmountable challenges, and to its ultimate and hard-won triumph, I hope you enjoy the journey and feel the spark of excitement for all the adventures into space yet to come.

And now . . . let's go to the Moon!

OPPOSITE: An artist's rendering of the roll-out of the Apollo 11 Saturn V from NASA's Vehicle Assembly Building (VAB) at the Kennedy Space Center, May 20, 1969. It was headed to the launch pad.

CHAPTER 1

"PROGRAM ALARM!"

"THE LUNAR LANDING OF THE ASTRONAUTS IS MORE THAN A STEP IN HISTORY; IT IS A STEP IN EVOLUTION."

—*New York Times* editorial, July 20, 1969

ON JULY 20, 1969, THE BLEAK EXPANSE of Mare Tranquillitatis—Sea of Tranquility—sat roasting in the sun, as it had for the past 4 billion years. The relatively flat basin was formed when basaltic lava flowed into the vast region during the violent formation of the Moon; it is one of the dark blobs you see when you gaze at the body. Roughly 540 miles (870 km) across, the region took shape about 600 million years after the initial formation of the Moon.

Utter silence filled the void. Bleached rubble overlay the basaltic plain, ranging in color from chalky gray to cocoa brown. The view to the horizon revealed a flat expanse, relieved only by a few low crater ridges. Due to the curvature of the surface, the edge of the horizon fell away quickly, just 1.5 miles (2.4 km) in the distance. With no atmosphere to moderate temperatures, the sunlit sides of the rocks and boulders littering the plain reached well over 200 degrees Fahrenheit (93°C), while the shadowed sides plunged to –250 (–157°C). The landscape had an oddly smooth, worn quality due to billions of years of bombardment by micrometeorites but had been otherwise unchanged since its formation. The Sea of Tranquility was a museum of the early solar system, 4 million millennia old.

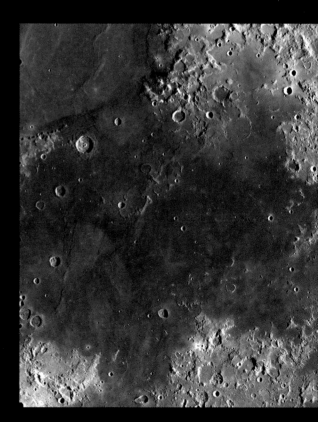

OPPOSITE: Neil Armstrong in the Lunar Module simulator, training for the Apollo 11 landing.

ABOVE: The Sea of Tranquility, landing site of Apollo 11, as seen from NASA's Lunar Reconnaissance Orbiter.

The *Eagle* at "pitchover," when Armstrong rotated the Lunar Module over from flying horizontally to flying "forward," so that they could see the lunar surface, far below.

But one change—the first major event in hundreds of millions of years—was about to occur. An infinitesimal occurrence in the history of the Moon, but a life-changing one for the small blue planet hanging in the sky, was imminent. High above and just over the lunar horizon, a tiny spacecraft had just ignited its rocket engine and was plunging toward the cratered expanse. Above that, a small companion spacecraft, crewed by a lone astronaut, orbited, awaiting the return of two brave explorers.

After nine years of crushing effort, Apollo 11 had arrived at the Moon.

"GO FOR POWERED DESCENT"

In the tiny aluminum craft rushing toward the surface, a scratchy radio transmission came through the headsets of the two astronauts standing inside, held in place by elastic straps connected to the floor. Neil Armstrong and Buzz Aldrin concentrated on their tasks—Armstrong, mission commander, monitored the view out the window as the Lunar Module (commonly abbreviated as LM, pronounced "lem") slowly orbited, and Aldrin, LM pilot, fussed with the radio controls, trying to improve reception.

The men could barely make out the instructions coming up from Earth because they were having difficulty keeping the small antenna on their spacecraft aimed at their home planet. Mike Collins, their companion in orbit high above, was relaying instructions as he could, and the most important of these was just coming in.

A garbled "*Eagle*, Houston. If you read, you're go for powered descent. Over," echoed through the metal interior of the LM. The voice was that of capsule communicator, or CAPCOM, Charlie Duke, another Apollo astronaut at Mission Control back in Houston, Texas. The CAPCOM served as the voice link between Mission Control and the astronauts in flight. Mike Collins, orbiting above in the Command Module

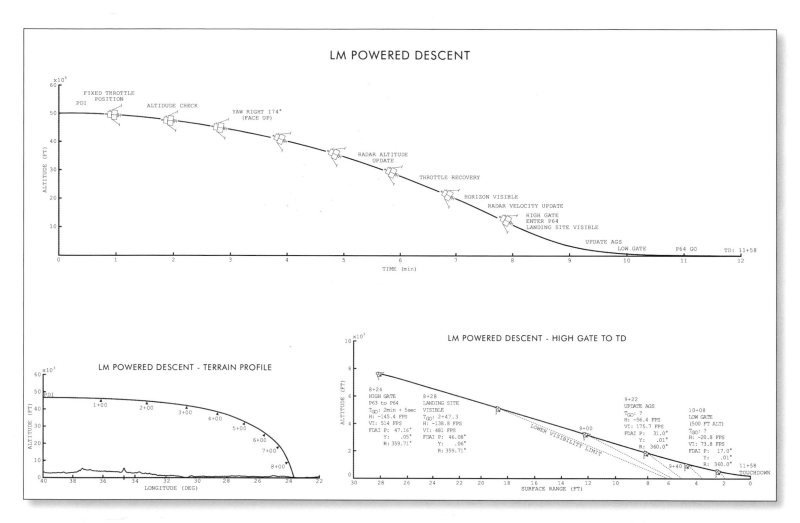

LM POWERED DESCENT

(Top graph — ALTITUDE (FT) ×10³ vs TIME (min))

- PDI
- FIXED THROTTLE POSITION
- ALTITUDE CHECK
- YAW RIGHT 174° (FACE UP)
- RADAR ALTITUDE UPDATE
- THROTTLE RECOVERY
- HORIZON VISIBLE
- RADAR VELOCITY UPDATE
- HIGH GATE / ENTER P64 / LANDING SITE VISIBLE
- UPDATE AGS
- LOW GATE
- P64 GO
- TD: 11+58

LM POWERED DESCENT - TERRAIN PROFILE

(ALTITUDE (FT) ×10³ vs LONGITUDE (DEG))

PDI, 1+00, 2+00, 3+00, 4+00, 5+00, 6+00, 7+00, 8+00

LM POWERED DESCENT - HIGH GATE TO TD

(ALTITUDE (FT) ×10³ vs SURFACE RANGE (FT))

8+24 HIGH GATE	8+28 LANDING SITE	9+22 UPDATE AGS	10+08 LOW GATE (500 FT ALT)
P63 to P64	VISIBLE		
T_{GO}: 2min + 5sec	T_{GO}: 2+47.3	T_{GO}: ?	T_{GO}: ?
H: -145.4 FPS	H: -138.8 FPS	H: -56.4 FPS	H: -20.8 FPS
VI: 514 FPS	VI: 481 FPS	VI: 175.7 FPS	VI: 73.8 FPS
FDAI P: 47.16°	FDAI P: 46.08°	FDAI P: 31.0°	FDAI P: 17.0°
Y: .05°	Y: .06°	Y: .01°	Y: .01°
R: 359.71°	R: 359.71°	R: 360.0°	R: 360.0°

LOWER VISIBILITY LIMIT

9+00, 9+40, 11+58 TOUCHDOWN

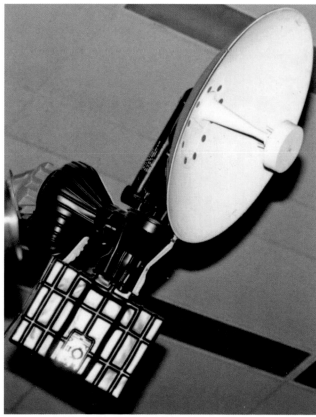

ABOVE: From the original *Apollo 11 Flight Plan*, the lunar landing profile.

LEFT: The S-band antenna on the Lunar Module was steerable to maintain communication with Earth and, if necessary, the Command Module. It was adjustable via Aldrin's control panel. There was also an omnidirectional antenna as a backup in case the S-band antenna lost contact.

(CM) *Columbia*, relayed the instruction: "*Eagle*, this is *Columbia*. They just gave you a go for powered descent." Aldrin confirmed receipt of the message.

This was it. They were about to be the first two humans to land on the Moon . . . or not.

It was the "or not" part that was vexing. For almost a decade, NASA (the National Aeronautics and Space Administration) and its contractors had been working full-out, in grinding overtime on multiple shifts, to bring its spaceflight efforts to this single moment: landing Americans on the Moon. But such a landing was far from assured; only twice before had Apollo spacecraft flown to the Moon, and neither had landed—Apollo 8 had orbited for about a day and come

ABOVE: A NASA-created rendering of what the Apollo 8 capsule would have looked like orbiting the Moon in 1968.

RIGHT: The view out of Aldrin's window in the Apollo 11 Lunar Module midway between the Earth and the Moon.

home, and Apollo 10 had flown a "dry run" in which the LM made a daring low-altitude sweep past the lunar surface before boosting back to the orbiting Apollo CM overhead. Apollo 11 would be the first actual landing attempt. So, as Gene Kranz, the flight director during the landing later said, "From this point forward, we would have one of three outcomes. We would land on the Moon, we would abort while trying to land, or we would crash. The last two were not good."[1]

Not good indeed. But the pair inside the LM were not thinking of alternative outcomes at this moment. They were intensely focused on getting down to the gray surface, which was slowly growing larger in the windows of their frail landing craft.

Their intensive training would soon pay major dividends. The highest compliment an Apollo astronaut could give to their massive support team back in Houston and the ground team in Florida at the Kennedy Space Center, where the rockets launched from, was to say, "The mission was just like the simulations." Indeed, they had simulated all phases of the Apollo 11 flight hundreds and hundreds of times. The simulation supervisors—known as Sim Sups—threw all

kinds of nasty problems at the astronauts to sweat out in the safety of Earth-bound simulators to see how they would do. To be specific, Armstrong had spent 959 hours in various kinds of training for this flight, and Aldrin 1,017 hours. About a third of that had been in LM simulators. So far, their experience over the Moon had earned that desired compliment—it was just like the simulations.

And then, suddenly, it wasn't.

LMA 790-1

PROJECT APOLLO

lem
LUNAR EXCURSION MODULE

FIRST MANNED LUNAR LANDING
FAMILIARIZATION MANUAL

GRUMMAN AIRCRAFT ENGINEERING CORPORATION • BETHPAGE, L. I., N. Y.

Cover from the 1965 *Lunar Excursion Module Familiarization Manual,* an early guide to basics of the lunar lander. As the Apollo program progressed, the delightfully dramatic nature of the illustrations on such documents waned.

"IT'S A 1202 . . ."

The first of many complications had just inserted itself into the landing—the radio transmissions from the LM dropped off the screens at Mission Control. The lost signal carried not just the voices of the astronauts, but also data from multiple onboard systems, which allowed controllers in Houston to monitor the landing. Without that data, they were unable to fully support the crew. Aldrin manually activated a secondary, omnidirectional antenna, which forced him to take his eyes off the all-important readouts from the onboard computer. This antenna was less accurate than the primary antenna, but it had a wider field of view, requiring less finesse to aim properly. The data link soon came back online.

The next problem took a bit longer to spot—Armstrong realized that they were about 3 miles (5 km)

off-target, flying ahead of where they wanted to be. This was due to a number of factors, but primarily, there had been some residual air in the docking tunnel of the CM, and when the LM uncoupled, it got a slight Champagne cork–like pop. That is all it took to throw them off course by miles. Armstrong said only, "Our position checks downrange show us to be a little long."

Moments later, CAPCOM Duke said, "Roger. You are go. You are go to continue powered descent. You are go to continue powered descent."

Then the data link to the ground dropped out again.

As Armstrong continued to maneuver the LM toward a proper orientation for landing, the fuel, which was now down to about half its original load, began to unexpectedly slosh around in the onboard tanks,

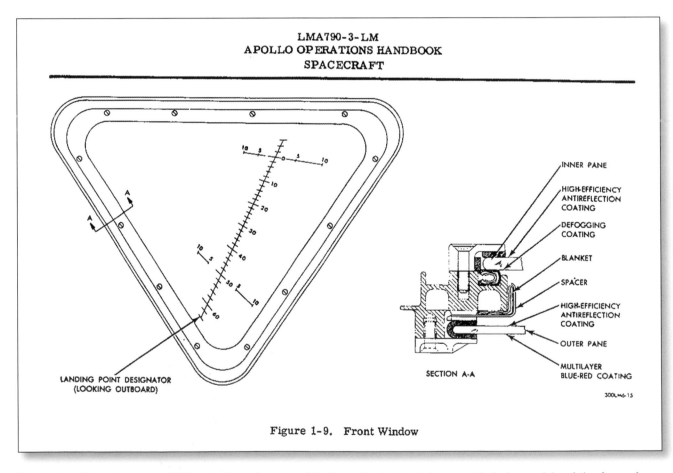

LMA790-3-LM
APOLLO OPERATIONS HANDBOOK
SPACECRAFT

LANDING POINT DESIGNATOR
(LOOKING OUTBOARD)

INNER PANE

HIGH-EFFICIENCY ANTIREFLECTION COATING

DEFOGGING COATING

BLANKET

SPACER

HIGH-EFFICIENCY ANTIREFLECTION COATING

OUTER PANE

MULTILAYER BLUE-RED COATING

SECTION A-A

300LM6-15

Figure 1-9. Front Window

The Landing Point Designator (LDP) etched into the glass of the Lunar Module's window provided a low-tech but fail-safe way for Armstrong to check the accuracy of his navigation to the lunar surface. From the *Lunar Module Operations Handbook*.

making the spacecraft wobbly and harder to control. Complication number three.

Then came a transmission that caused many hearts to skip a beat in Mission Control. In a cool but urgent tone, Armstrong said, simply,

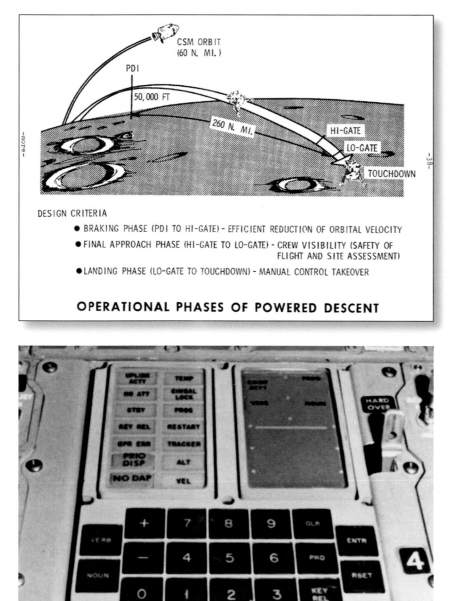

DESIGN CRITERIA

● BRAKING PHASE (PDI TO HI-GATE) - EFFICIENT REDUCTION OF ORBITAL VELOCITY
● FINAL APPROACH PHASE (HI-GATE TO LO-GATE) - CREW VISIBILITY (SAFETY OF FLIGHT AND SITE ASSESSMENT)
● LANDING PHASE (LO-GATE TO TOUCHDOWN) - MANUAL CONTROL TAKEOVER

OPERATIONAL PHASES OF POWERED DESCENT

TOP: Armstrong and Aldrin were approximately at the point indicated in this Apollo 11 press kit rendering from 1969 when the Apollo Guidance Computer locked up with the 1202 error.

ABOVE: The Lunar Module's digital computer, the Apollo Guidance Computer. In mid-landing, the screen to the right suddenly displayed the 1202 alarm . . . and little else.

"Program alarm." The onboard computer, critical to a successful landing, had been steadily displaying ever-changing readouts of their orientation and altitude. It now locked up, displaying four unforgiving green digits: 1202. The altitude and rate indicators were frozen, and the 1202 error code appeared to indicate that something was wrong with the computer or its programming. This meant the possibility of having to abort the landing—a dangerous operation at any time.

With thinly veiled impatience, Armstrong followed with, "Give us a reading on the 1202 program alarm." They were at about 33,000 feet (10,000 m) and descending fast.

In Mission Control, there were blank stares on some faces, while other controllers flipped rapidly through their notes—nobody knew immediately what a 1202 alarm was, given the intense complexity of the Apollo Guidance Computer's (AGC) programming.

The descent was at an impasse. As Aldrin later said, "In simulations, someone's training you to give a certain response. So you want to do the right thing in the simulation. When it's not a simulation, you want to do the right thing to get the mission done."[2]

Armstrong summed it up more succinctly: "We had gone that far and we wanted to land. We didn't want to practice aborts."[3]

The landing of the first humans on the Moon hung in a precarious balance.

CHAPTER 2

RACE TO THE MOON

"WE CHOOSE TO GO TO THE MOON IN THIS DECADE AND
DO THE OTHER THINGS, NOT BECAUSE THEY ARE EASY, BUT
BECAUSE THEY ARE HARD . . ."

—John F. Kennedy,
speech at Rice University, September 1962

AMERICA'S JOURNEY TO THE MOON did not start out on a promising note, despite the fact that the nation's technological prowess in the preceding decades seemed to indicate that such an undertaking would be an assured success.

Since the Allied victory in World War II, in which US dominance in technology and industrial capability had quite literally won the day, the country had considered itself the frontrunner in innovation. The Soviet Union emerged from that conflict a battered, ruined nation, barely able to feed its starving population. Europe, which was divided between the Western allies and Soviet-dominated nations, was also shattered by the effects of war and was slowly rebuilding itself. Japan was recovering from the devastation of nearly a decade of war and twin atomic bombings. Of the combatants and their associated states, only the US, exhausted yet physically untouched by years of warfare, emerged relatively unscathed. By the late 1950s, there was a new car in most driveways, 50 million televisions had sprouted in American living rooms, and modern appliances eased domestic drudgery in most kitchens.

OPPOSITE: Wernher von Braun's V-2 rocket, launching from Germany in a 1945 British test after the war.

ABOVE: Ground zero of a V-2 impact in London, 1945.

The economy was humming along to a peak in the 1960s. Life was good in America.

In the meantime, the Soviets were rebuilding furiously, drawing upon any possible advantage to do so. By the middle of the 1950s, they had managed to manufacture not just an atomic bomb but also, and far more worrying to the West, a hydrogen bomb, liberally aided by years of espionage that spirited away secrets from the US. And while their economy was not robust by capitalist standards, they were emerging as the second global superpower—a challenge to the US. This was not just in military terms, but soon also in technology and science. By the late 1950s, in no area was this challenge felt more keenly than spaceflight.

While scientific achievement

A V-2 rocket engine, retrieved from an impact site in London, 1944.

was a goal in space research, it was largely window dressing for the ongoing quest for military power. In the aftermath of World War II, both nations had built their nuclear arsenals to frightening levels with increasingly modern technology, but the delivering of incandescent destruction to their enemies still depended on bomber aircraft. These lumbering airplanes would have to fly across thousands of miles of heavily defended terrain (and, in the case of US-based targets, open ocean) to drop their deadly payloads. Enormous sums were spent building better bombers, but by the early 1950s, rockets that could deliver nuclear destruction to the enemy were being aggressively pursued. Once launched, they could not be stopped. Both countries poured money into the development of intercontinental ballistic missiles (ICBMs), and this was the real genesis of space exploration.

The earliest missile effort was the V-2 rocket, brainchild of Wernher von Braun. Von Braun was a prodigy in German rocketry. He joined the Nazi party in 1937, at age twenty-five, because he saw it

as the only way to continue his pursuit of rocketry. As the country headed toward war, he was recruited to head the German army's rocket program in Peenemunde. His mandate? To develop the first guided ballistic missile in history; one that could attack targets in England, Holland, Belgium, and other adjacent nations. By midwar, London, Antwerp, and other European capitals were feeling the brunt of von Braun's ingenious yet fearsome design.

As Germany slid into defeat in 1945, von Braun realized that he had a choice to make: he could surrender either to the advancing Russians or to the American army. Knowing of the depredations inflicted upon the Soviets by the German army, von Braun chose the US. After turning himself over to the US Army, von Braun and hundreds of his associates were taken out of Europe and relocated to America along with much of their missile-making hardware. When the Soviets invaded Germany, they captured whatever—and whomever—the Allies had left behind. Thus was born the rocketry efforts of both superpowers.

A SHOCK TO THE WEST

In America, von Braun continued to refine his designs, and the Redstone missile that carried the first flights of the Mercury program was developed from the V-2. In the USSR, German and Russian engineers perfected the R-7 rocket—a significant departure from the V-2 design, but with similar roots. Both were designed to deliver nuclear warheads, and variants of each ultimately flew the earliest satellite launches. The Soviets launched their first satellite, *Sputnik*, on October 4, 1957.

This was a shock to the West and a profound wake-up call in the competition for achievement in spaceflight. Western governments had known that the Russians were working on satellites and missiles, but they were caught flat-footed when the first orbital satellite launch occurred. While *Sputnik* was just a 2-foot-wide (61-cm), 184-pound (83-kg) sphere that did little more than transmit an electronic beep to the

world, the launch of *Sputnik 2*, which followed less than a month later, was much more alarming. It carried a 13-foot-long (4-m), 1,100-pound (500-kg) capsule with a live dog inside. While technically impressive overall, this payload also approximated the weight of a nuclear warhead—one capable of being dropped anywhere on the United States from orbit. There would be no stopping such a weapon.

In the US, the navy had been tasked with launching America's first satellite. Project Vanguard would use a purpose-built rocket, in part to show the world that this was a purely peaceful undertaking that was not being launched on a repurposed missile. Both the Vanguard rocket and satellite were far smaller than their Russian counterparts—the Vanguard satellite was about the size of a grapefruit. When a launch was attempted in December 1957, a full two months after the success of *Sputnik*, it failed spectacularly on live

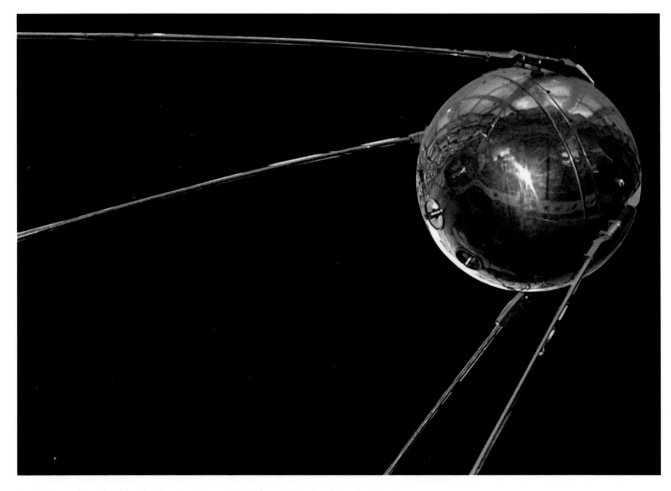

Sputnik was launched by the Soviet Union in October 1957, shocking the Western powers.

12

OPPOSITE: The failure of the US Navy's Vanguard Test Vehicle 3 (TV3 for short) launch attempt in December 1957 was televised to the world.

ABOVE: From left to right, William Pickering, James Van Allen, and Wernher von Braun hold up a replica of Explorer 1 at a press conference after the launch of America's first satellite on January 28, 1958.

television. The world watched as the rocket exploded on the pad and the tiny satellite toppled off the top of the booster, falling through the fireball and rolling across the tarmac. It was ultimately found jammed under a dumpster, still beeping away.

Between the Soviet success and the Vanguard failure—variously labeled by the media as "Flopnik," "Kaputnik," and "Stayputnik"—the US government became increasingly distressed. Officials had not wanted to use von Braun for a public role in the American space program, in part due to his Nazi past, but with the Soviet triumph, all bets were off. When asked what he could accomplish after the Vanguard failure, a frustrated von Braun stated that he could get a satellite in orbit "in ninety days." Desperate after "Flopnik,"

the government turned him loose on the project, and Explorer 1 was launched in less than sixty days—atop a repurposed Redstone missile—on January 28, 1958. With a successful US satellite launch, the space race went into high gear.

NASA was created later that year, officially making US efforts at spaceflight a civilian undertaking, with the air force continuing to develop nuclear missiles. In truth, the two remained blended for much of the 1960s, with NASA missions flying on air force rockets until 1966. In contrast, the Soviets kept their human spaceflight program under tight military control.

This Soviet program continued to produce one impressive spaceflight first after another: the first orbital satellite, the first living thing in space (the dog Laika on *Sputnik 2*), the first spacecraft to fly past the Moon's orbit, and more. This was profoundly embarrassing to the United States, which steadfastly denied that it was in a race with the Soviets but was nonetheless painfully aware of being seen as second best by both domestic and international media. Perception was important during the Cold War, and a second-class technological and military power would lose support in the international community. That support—especially from nonaligned countries deciding whether to follow the doctrine of Communism or Democracy—was deemed critical by US leaders. In a bifurcated world where most countries would have to decide on alignment with one superpower or the other, the United States could not afford to lose influence.

SOVIET MEN, AMERICAN CHIMPS

The embarrassments were not over yet. By 1960, both countries were racing to put the first human in space. The Soviets were working on their R-7 and the Americans on von Braun's far smaller Redstone, and later the larger and more powerful Atlas rocket. Both programs were moving along at a rapid pace, with NASA flying a chimpanzee named Ham (honoring the Holloman Aerospace Medical Center, an air force clinic in New Mexico) on January 31, 1961, aboard the new Project Mercury space capsule. There were concerns over how spaceflight might affect humans, and flying a primate first allowed doctors to study its responses to weightlessness. This was merely a sixteen minute sub-orbital flight, since the Redstone booster being used was not capable of propelling the capsule into Earth orbit. Ham returned agitated but unharmed, and preparations went into overdrive to launch the first US astronaut into space.

Then came the ultimate in one-upmanship. On April 12, 1961, the Soviets launched Vostok 1, with cosmonaut Yuri Gagarin aboard. He made a single orbit of the Earth and then reentered, landing in Siberia 108 minutes after launch. The world celebrated the achievement, with the US government reluctantly sending a congratulatory message to the Soviet leadership, even while nursing its wounded pride. Once again, the USSR had outshone America's finest.

NASA finally launched a manned Mercury flight on May 5, 1961. This was another suborbital jaunt,

just over fifteen minutes in duration, but it showed that the Mercury spacecraft could be trusted with a human passenger. America finally had its "man in space." The lone passenger was Alan Shepard, the first pick among the original seven Mercury astronauts chosen in 1959. Despite a ticker-tape parade in New York City and a presidential medal, the achievement rang hollow. The Soviet Union still held all the records. It had flown two more dogs into orbit and returned them to Earth alive in 1960, launched the first spacecraft to Venus in 1961 (though its radio failed before the flyby, it was still a notable accomplishment), put the first human into orbit, and would continue to chalk up spaceflight firsts for many more years.

This was galling to US leadership in general, and specifically to the newly elected president, John F. Kennedy, who entered the office in January

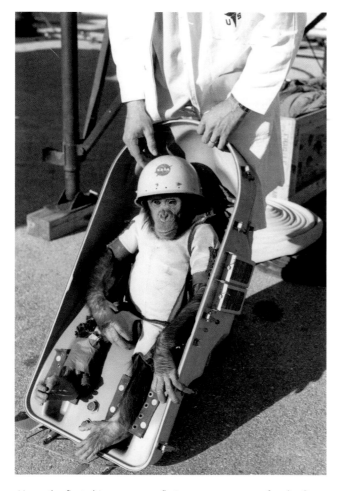

Ham, the first chimpanzee to fly in space, prepares for the first "crewed" Mercury flight.

Alan Shepard became America's first human in space, launched aboard the first manned Mercury capsule on May 5, 1961.

1961. Frustrated with the Soviet Union's ascendancy in both orbit and world opinion, Kennedy tasked his vice president, Lyndon B. Johnson, to poll leaders in NASA, the aerospace industry, and the military to ascertain what goal the United States might attempt that would assure it of a "win" in the space race.

With the results of this poll in hand, Kennedy gathered his advisors and a few key NASA chiefs to plan something audacious in an attempt to burnish America's technological image. It would not be an easy task. Johnson had reported back that there was precious little to crow about in domestic spaceflight achievements, but with a strong, committed effort, the US might be able to beat the Soviets to a Moon landing. Anything less ambitious, such as a crewed orbiting station or manned lunar flyby, could result in another second-place finish, or at best, a draw. Neither country yet had a rocket capable of delivering the required mass to the Moon's surface for a manned landing, nor had either succeeded at navigating any machine successfully to a target beyond Earth orbit. Likewise, neither power had yet landed anything on another world. A manned lunar landing appeared to be the best technological challenge that would assure the US a firm "win" in space.

From a memo prepared for LBJ by NASA administrator James Webb and Secretary of Defense Robert McNamara, in May 1961:

> We recommend that our National Space Plan include the objective of manned lunar exploration before the end of this decade. It is our belief that manned exploration to the vicinity of and on the surface of the Moon represents a major area in which international competition for achievement in space will be conducted. The orbiting of machines is not the same as the orbiting or landing of a man. It is man, not merely machines, in space that captures the imagination of the world.[4]

The summary from a letter to LBJ from Wernher von Braun, who by that time was widely regarded as the most capable rocketry expert in the free world, was even more urgent in tone: "Summing up, I would like

Jerome Wiesner, Kennedy's science advisor, was lukewarm on making the lunar landing program the sole focus of NASA's efforts through the 1960s.

to say that in the space race we are competing with a determined opponent whose peacetime economy is on a wartime footing. Most of our procedures are designed for orderly, peacetime conditions. I do not believe that we can win this race unless we take at least some measures which thus far have been considered acceptable only in times of national emergency."[5]

In the following weeks, Kennedy discussed the lunar landing idea with his trusted advisors. Many were for it; some, including Jerome Wiesner, Kennedy's science advisor, did not want the entirety of the US's national space effort focused on a lunar landing program. But while his voice was an important one, he was overruled, and on May 25, 1961, with only fifteen minutes of Alan Shepard's suborbital Mercury flight to back him up, Kennedy took an audacious gamble in his address to the US Congress.

A MAN ON THE MOON

In a speech called Special Message to the Congress on Urgent National Needs, which outlined a number of national imperatives, Kennedy waited until the end to drop the largest bombshell of the second half of the twentieth century:

> Space is open to us now; and our eagerness to share its meaning is not governed by the efforts of others. We go into space because whatever mankind must undertake, free men must fully share.

I therefore ask the Congress, above and beyond the increases I have earlier requested for space activities, to provide the funds which are needed to meet the following national goals:

First, I believe that this nation should commit itself to achieving the goal, before this decade is out, of landing a man on the Moon and returning him safely to the Earth. No single space project in this period will be more impressive to mankind, or more important for the long-range exploration of space;

President John F. Kennedy delivers his Special Message to the Congress on Urgent National Needs speech on May 25, 1961.

and none will be so difficult or expensive to accomplish. We propose to accelerate the development of the appropriate lunar space craft. We propose to develop alternate liquid and solid fuel boosters, much larger than any now being developed, until certain which is superior. We propose additional funds for other engine development and for unmanned explorations—explorations which are particularly important for one purpose which this nation will never overlook: the survival of the man who first makes this daring flight. But in a very real sense, it will not be one man going to the Moon—if we make this judgment affirmatively, it will be an entire nation, for all of us must work to put him there.[6]

The speech went on for a few more minutes, studded with applause and nods from the Democrats in attendance; many others appeared to be stunned. But the core message was clear: the president was committing the nation to a vast and bold effort in space.

His more famous speech on the US's lunar ambitions followed in September 1962 at Rice University in Texas, where he addressed a stadium packed with attendees sweltering in the hot, late-summer sun. This was the second half of his sales pitch—this time to the American public at large. The most memorable phrases are from midspeech:

> But why, some say, the Moon? Why choose this as our goal? And they may well ask, why climb the highest mountain? Why, thirty-five years ago, fly the Atlantic? Why does Rice play Texas?
>
> We choose to go to the Moon! We choose to go to the Moon in this decade and do the other things, not because they are easy, but because they are hard; because *that* goal will serve to organize and measure the best of our energies and skills, because that challenge is one that we are willing to accept, one we are unwilling to postpone, and one we intend to win.

JFK gives his second "Moon speech" at Rice University in September 1962, effectively securing the "deal" with the American public for the Apollo program.

And with that, the deal was done. The cost would be staggering, the effort exhausting, the commitment total. There were naysayers and pushback—this would reshuffle the priorities of an entire nation and affect its citizenry deeply. It was the biggest technological undertaking since the Manhattan Project of World War II, and even that massive effort would pale in comparison in both cost and time. But the forward momentum was irresistible, and enough of the country was behind it to forge ahead.

America was headed to the Moon.

MEN FOR THE MOON

"AMERICA ALWAYS DOES BEST WHEN IT ACCEPTS A CHALLENGING MISSION. WE INVENT WELL UNDER PRESSURE."

—Buzz Aldrin, Apollo 11 astronaut

WHEN KENNEDY MADE HIS FIRST announcement of a massive national effort to land Americans on the Moon, Neil Armstrong was in Seattle. He was not yet a NASA astronaut; in fact, there were only the seven Mercury astronauts on the NASA roster at the time. The second group—dubbed the "New Nine"—would not be announced until September of 1961, and Armstrong would be part of that cadre.

There would be multiple astronaut selections after this group, and Aldrin and Collins would be invited to join group three. Kennedy had stated his goal to reach the Moon and that declaration was heard across the country from from the Navy's Test Pilot School at Patuxen River, Florida, to the tarpaper shacks of Edwards Air Force Base in the high desert of California, where the best test pilots labored, wringing the bugs out of dangerous and untried fighter planes. In both places, test pilots responded eagerly to the call.

OPPOSITE: The Apollo 11 crew—Buzz Aldrin, Neil Armstrong, and Mike Collins (left to right)—relax before one of the many training exercises they did leading up to the mission.

RIGHT: Neil Armstrong flew multiple missions over wartime Korea in the early 1950s in a Grumman F9F Panther. He can be seen in the background plane in this image.

NEIL A. ARMSTRONG: THE QUIET ONE

Neil Armstrong had spent time in the navy, joining in 1949 and attending navy flight school. By 1951 he was flying sorties over wartime Korea, with pitched battles raging not far below him. He flew seventy-eight missions there, experiencing his first brush with death when bailing out of a crippled fighter. It would not be his last.

After leaving the navy, Armstrong attended Purdue, receiving a BS in aeronautical engineering. After a brief stint at the Lewis Flight Propulsion Laboratory in Cleveland, Ohio (which would later become a NASA facility), he was sent to Edwards Air Force Base to be trained as a test pilot. Edwards was the beating heart of cutting-edge flight tests and was where the hotshot pilots wanted to be. While Armstrong never thought of himself as a hotshot, he knew it was where the action was in high-performance aircraft.

At Edwards, Armstrong was already in a form of astronaut training—he was working for NASA's civilian predecessor, the National Advisory Committee for Aeronautics (NACA), flying the X-15 rocket plane and preparing to transition to an ultimately cancelled air force spaceflight program called Project Dyna-Soar. He was one of only two future NASA astronauts to fly the X-15 (Joe Engle, who later flew space shuttles, was the other). Armstrong had earlier been recruited to a short-lived air force program called Man in Space Soonest (with the unfortunate acronym of MISS), which had been part of the national effort to beat the Russians to putting a human in space. When NASA absorbed MISS into its own Mercury program, the air force transitioned to designing a spaceplane with military potential—this was Project Dyna-Soar. It would launch atop a rocket (the X-15 launched from underneath the wing of a B-52 bomber), taking two air force astronauts into orbit to perform military tasks ranging from biomedical research to spying on the Russians, then reenter as a winged craft—much as the space shuttle would later do—to be refurbished and reused.

Armstrong has said that he does not recall much of his immediate reaction to Kennedy's announcement to commit NASA to a lunar effort, other than excitement about the new technologies that would be involved. But

Neil Armstrong in 1956.

when he heard that NASA was looking to expand the astronaut corps, he applied for a slot. Though his application arrived at NASA a full week after the deadline, an associate of his from Edwards, now working at NASA, spotted it and surreptitiously slipped the application into the active pile for consideration.[7] Armstrong had been flying the X-15 for two years by then.

In September of 1962, NASA's Deke Slayton called Armstrong and asked if he would like to join the newly expanded astronaut corps, to which Armstrong replied with an unqualified yes. Slayton had been selected as one of the original Mercury astronauts but was later found to have a heart murmur and grounded. He was subsequently made the coordinator of astronaut activities. It was in this capacity that he not only informed new astronauts that they had been selected but later, as the director of flight crew operations, he also became the person who determined flight assignments, making him a very powerful man in the astronaut community.

Armstrong poses with the X-15 at Edwards Air Force Base.

Armstrong joined NASA that month as its first civilian astronaut. He soon began training for his first spaceflight for the Gemini program, the two-seater predecessor to Apollo. Gemini would be not only the test bed for the technologies and techniques required to move ahead in the lunar landing program, but also a training ground for many of the astronauts who would fly during Apollo.

"NASA felt that its new astronauts with little experience with the sophistications of orbital mechanics or the differences between aircraft and spacecraft needed a quick primer," Armstrong later said. Of these complexities, he noted, "Some of them were new to me, but overall I didn't find the academic burden to be overly difficult."[8]

But that was the classroom work. Soon the New Nine would find themselves on the road, enduring grueling, cross-country schedules. First, they were sent to familiarize themselves with NASA's new facilities—then under construction across the US—and then to the aerospace contractors that would be building the Gemini and Apollo hardware. The schedule was grinding, with much of their time spent in flights between Houston (where they were based) and Florida (where the Kennedy Space Center was being built), to Huntsville, Alabama, where the Marshall Space Flight Center was building rocket boosters, and on to St. Louis, where McDonnell Aircraft was beginning work on the new Mercury capsule. Then they might fly to Southern California, home of aerospace contractors Lockheed, North American Aviation, and the Douglas Aircraft Company. The workload extended well beyond normal office hours and would be just a foretaste of things to come.

EDWIN E. "BUZZ" ALDRIN: "DOCTOR RENDEZVOUS"

Buzz Aldrin followed a very different path into NASA's astronaut corps. After graduating from West Point in 1951, he was commissioned as a second lieutenant in the air force and deployed to Korea, where he flew sixty-six combat missions, shooting down two enemy aircraft. He spent time in Europe and at Nellis Air Force Base in Nevada, and he then went on to study at Massachusetts Institute of Technology (MIT) in 1959.

It was while attending MIT that Aldrin watched Kennedy's address to Congress on a small black-and-white television at his home in Cambridge. He was excited by the loud applause and standing ovation by the congressmen in attendance. Just a few days prior, Aldrin had received a letter from Ed White, a classmate at West Point and fellow air force pilot who would soon fly on Gemini and mentioned that he intended to apply for the second group of astronauts. Unlike White, Aldrin had not attended military test-flight school, a requirement for application. Downcast but determined, he decided to finish his PhD program at MIT before applying to be an astronaut.

Aldrin recalls:

> The country was swept up in the space program, and I wanted to be a part of it. But NASA retained its requirement that astronauts have a diploma from a military test pilot school—not one of my credentials. Since I knew that the Moon landing program Kennedy had described would need astronauts with skills other than the ones they drummed into you at test pilot school, I opted for another eighteen months of intensive work on a doctorate in astronautics, specializing in manned orbital rendezvous."[9]

Aldrin knew there would be a need for multiple spacecraft to rendezvous in Earth orbit (and, as it turned out, lunar orbit) to attain the goal of a lunar landing. He also knew that his work at MIT would make him unique in the astronaut corps. He also suspected that the computers being planned for such

Edwin "Buzz" Aldrin while in the air force.

maneuvers, which did not yet exist, would be fallible—once developed, they would have to be miniaturized to fly in spacecraft, and small electronics often failed in that era. Someone with experience in orbital rendezvous could quite literally save the day in such an emergency, and by pure serendipity, his hunch would later be proved correct during his Gemini 12 flight.

Upon receiving his doctorate degree, Aldrin applied to NASA to become an astronaut, but as he suspected he would be, he was turned down due to lack of test-pilot experience. In January 1963, he took a new assignment at the Los Angeles Air Force Base in southern California with the office responsible for planning orbital rendezvous techniques for the new Gemini program. As he was preparing for the new job, NASA changed its rules regarding astronaut selection, and he applied to become an astronaut again just as he was moving into his new job. That September he received the much-anticipated call from Deke Slayton. "We'd sure like you to become an astronaut," he recalls Slayton saying.

Aldrin of course said yes immediately—"Shoot Deke, I'd be delighted to accept."[10] Upon arrival, Aldrin dug in to his specialization—designing orbital rendezvous techniques for the Gemini missions.

MICHAEL COLLINS: THE RACONTEUR

Mike Collins was also a product of West Point, having graduated in 1952. Given a choice between the army and the air force, Collins chose the air force. Unlike his crewmates in Apollo 11, Collins did not fly in the Korean War, serving instead in Europe through the mid-1950s. After this military service, he applied to test-pilot school at Edwards Air Force Base though he felt it was a long shot since he had only the minimum number of flying hours required—1,500 in total. Despite the odds, he was accepted and reported to Edwards in August 1960.

While at Edwards, Collins was thrilled when he saw the news coverage of the first American to orbit the Earth, John Glenn, in February 1962. Like Armstrong, Collins applied for the second group of NASA astronauts. Though he was not accepted, he applied again at the next opportunity and soon received "the call" from Slayton. He was brought in as a member of the third group along with Aldrin.

Collins remembers that the early training included 240 hours of instruction on spaceflight, including 58 hours of geology training, which did not thrill him.[11] At the time, he never thought that he would spend nearly a full day staring down at geological formations from lunar orbit while his crewmates explored its surface. The geology coursework just seemed like a waste of time.

Like the others, he was assigned to an area of specialization to support the Gemini program, and he picked space suits and EVA (extra vehicular activity, or spacewalks) as his concentration, working as a go-between for NASA and its contractors. At the time this seemed a reasonable choice, but as he moved deeper into his concentration on space suits, he realized that this was a life-or-death specialization—a few layers of fabric and rubber separated a person from the harsh vacuum of space. He later said, "I was trying to guess

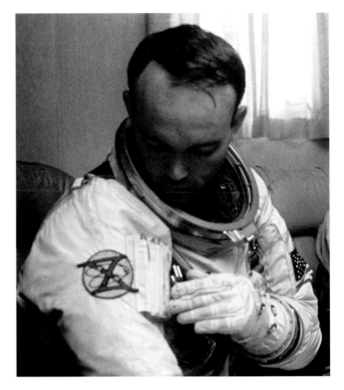

Mike Collins prepares for his flight in Gemini 10, July, 1966.

years in the future what thirty people, myself plus the other twenty-nine in the office [would be doing]—I was making commitments for that group of thirty people for actions we would or would not take a couple of years in the future, and that's kind of a scary responsibility."[12]

As the trio labored within the ranks of the Astronaut Office, none knew what lay in store for them. Oh sure, they were part of the great adventure of the American space program, but five of the original seven astronauts from the Mercury program remained in flight status. Slayton had been sidelined by heart problems, and now Al Shepard, the first among them to fly the Mercury capsule, had been diagnosed with a rare inner-ear disorder that grounded him. But that left five guys who would almost certainly be first in line for the Gemini flights, and so for Armstrong, Aldrin, and Collins, assignment to a plum mission was an uncertainty.

Nonetheless, they threw themselves into their assigned tasks, traveling wherever they were instructed and doing whatever was needed. But soon the flight assignments would come, and for each of them, these would be life changing.

THIS IMAGE WAS CREATED from a series of photos taken out the window of the Lunar Module after the Moonwalk as the astronauts were using up the rest of the film in the magazines before discarding the cameras outside the LM. The original images were not released to the press at the time of the mission, due to the fact that NASA only chose the most "press-worthy" pictures to send out. Many photos, including multiple-image montages such as this one, sat buried in NASA's archives for decades until a release was orchestrated by NASA in 2015. Some are marvelous images of the landing site, while others are out-of-focus or tilted accidental shutter releases caused by the bulky gloves of the Moonsuits while carrying the cameras.

Seen here is the site of the Moonwalk after the astronauts had reentered the Lunar Module. Note the footprints crossing the nearby soil—footprints that will last millions of years, only slowly being softened by a constant rain of micrometeorite impacts. The flag is still upright at center, but it will topple a few hours later when the ascent stage departs the Moon in a sudden blast of thrust. To the lower left is the shadow of the Lunar Module, and to the right the Reaction Control System (RCS) "quad" can be seen—an assemblage of four thrusters used to steer the spacecraft.

HOW DO WE DO THIS?

"DO WE WANT TO GET TO THE MOON OR NOT?"

—John Houbolt, engineer at
NASA Langley Research Center

KENNEDY'S ANNOUNCEMENT OF America's intention to land humans on the Moon may have stirred a nation, but the idea was hardly new. Alone in our sky, the Moon was the one place we could see in detail—the mountains, rilles, and craters were there to survey for anyone with a telescope of moderate size. Even a pair of binoculars or small handheld telescope revealed breathtaking detail. Dreamers and visionaries had been thinking about making the journey for centuries.

Perhaps the most famous early musings are contained in Jules Verne's novel *De la terre à la lune* (*From the Earth to the Moon*), first published in 1865. In this prescient book, Verne told of a journey undertaken by a small band of adventurers and supported by the Baltimore Gun Club. The club president decides that a large craft, patterned after a giant artillery shell, could be fired from Earth and reach the Moon, given an equally large cannon. With about $6 million in funds raised from US and European sources, the huge gun is built in Tampa Town, Florida—not far from the current location of the Kennedy Space Center—in a 900 × 60–foot

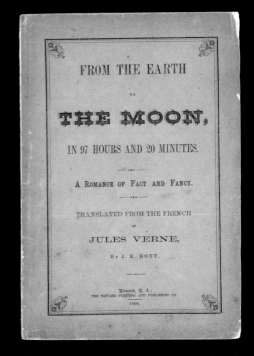

OPPOSITE: Russian scientist Konstantin Tsiolkovsky theorized that traveling to the Moon via shells shot from a giant cannon would kill the human crew instantly. His preferred method of locomotion through space was rockets—and he was proved correct.

ABOVE: An illustration from Jules Verne's 1865 novel *From the Earth to the Moon* showing a train of spacecraft shot from a giant cannon and headed to the Moon.

toward the Moon, accompanied by chickens and two dogs. Along the way they experience weightlessness and the stunning sight of Earth's nearest neighbor.

The use of ballistic means (e.g., a giant cannon) for sending humans to the Moon was later dismissed by Russian scientist Konstantin Tsiolkovsky, who would become known as the father of Russian rocketry, when he estimated that the passengers would be subjected to over 20,000 times Earth's gravity when the gun was fired. But his argument was against methodology, not the notion of lunar travel. He developed early calculations that demonstrated how rockets, fueled with hydrogen and oxygen—the same fuels used by many modern rockets—could travel into and through space.

Tsiolkovsky, along with other visionaries such as American physicist Robert Goddard and German scientist Hermann Oberth, provided inspiration for many youthful minds pondering voyages through the cosmos, and one of them was a young Prussian named Wernher von Braun. Von Braun was active in German amateur rocketry in the early 1930s and, after committing his energies to the Nazi government before World War II, designed the V-2 ballistic missile. But von Braun's true goal was not to build instruments of war—though he did so with little apparent reservation—but to build machines to explore space.

His work, and that of others, informed and inspired rocket development after the war and into the late 1950s. In the United States, two engineers, Maxime Faget and Owen Maynard, were fascinated by these concepts and blended them with work being undertaken in the design of blunt-body reentry craft (what would later become known as "space capsules," the "blunt body" denoting the blunt, curved heat-shield reentry design). By 1960, these ideas had been combined with the rocket designs of von Braun, who by then was working hard on America's first powerful civilian booster, the Saturn I.

These booster and spacecraft technologies were then fused into what would become America's manned

Even a telescope of moderate size reveals astonishing details on the Moon.

ABOVE: Mercury capsules being constructed at McDonnell Aircraft, 1960.

LEFT: Designer of the Mercury, Gemini, and Apollo capsules, Maxime Faget.

space program. While the blunt-body design would find its first useful application in Project Mercury, the overall concept was being applied to a program that was, just that year, dubbed Apollo. How an astronaut (or astronauts) would land on the Moon and return to Earth was still under discussion, but the general framework had been laid out. In the meantime, McDonnell Aircraft was under contract to NASA to build the Mercury capsule that would fly the first Americans into orbit. The next steps toward the more ambitious lunar goal were still in flux.

Then came Kennedy's first Moon speech in May 1961, and the die was cast. NASA, under James Webb, was charged with meeting the goal, but just *how* this would be accomplished was still undetermined. NASA's small, back-burner, lightly funded program to someday get American astronauts to the Moon suddenly had a deadline of nine years, soon to be accompanied by a vast budget increase, and all bets were off—the Apollo program had to begin *now*.

DIRECT ASCENT VS. RENDEZVOUS IN SPACE

At that moment, the Apollo mission profile suggested a method called Direct Ascent to land on the Moon. This would involve a single, large spacecraft—one huge rocket that would launch one enormous lander that would fly to the Moon, land, then return to Earth without shedding any stages along the way. The method was deemed the simplest way to go about the

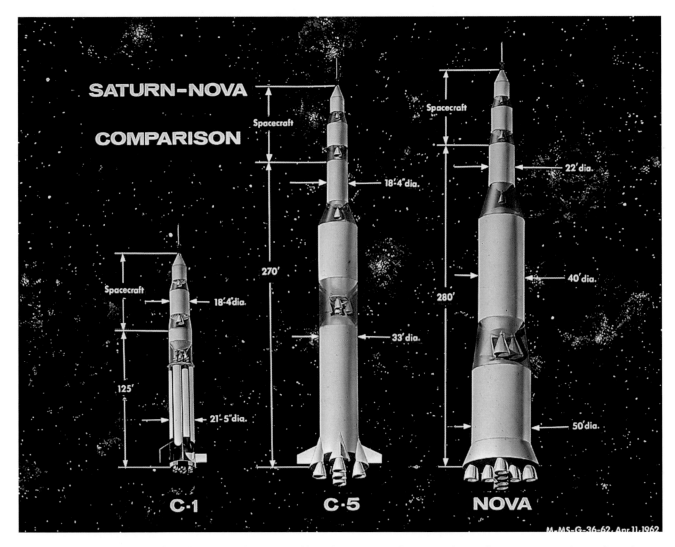

SATURN-NOVA COMPARISON

Spacecraft

18'-4"dia.

270'

33'dia.

C-5

Spacecraft

18'-4"dia.

125'

21'-5"dia.

C-1

Spacecraft

22'dia.

40'dia.

280'

50'dia.

NOVA

M-MS-G-36-62, Apr.11.1962

OPPOSITE: A Saturn I rocket launch in November 1962. Although von Braun's largest rocket at the time, it was a far cry from the Saturn V and could not even lift a complete Apollo spacecraft, as then envisioned, into orbit.

ABOVE: Comparison of the mammoth Nova rocket, to right, with the ultimate Moon rocket, the Saturn V, at center. At the time the Nova was being considered, the largest rocket in NASA's arsenal was the C-1, at left. Period illustration.

mission, as it did not require rendezvous and docking in space—something not yet attempted by NASA. It would, however, require building a huge booster that would have to be much more powerful than anything that existed at the time. This proposed rocket was known as the Nova.

The other method being explored, and one that von Braun favored as time went by, was called Earth Orbit Rendezvous, or EOR. This approach would require multiple flights of the Saturn I (this entire rocket only had about one-fifth the thrust of the future Saturn V that would ultimately deliver Apollo 11 to the Moon) to assemble a modular Moonship in orbit, and then send

it directly to the lunar surface. The other downside of EOR was that it would mean rendezvous, docking, and assembling some of the Moonship's components in space, and it was not yet known if this could be accomplished in practice. Despite the many challenges facing EOR, however, von Braun was undaunted—the moon landing was a national priority.

In either case, the Apollo spacecraft—consisting of a conical capsule atop a propulsion and landing stage with legs—would land, tail first, on the Moon. It would serve as the transport ship and lander, support the astronauts while there, and then blast off from the lunar surface and return them home. Flying Apollo

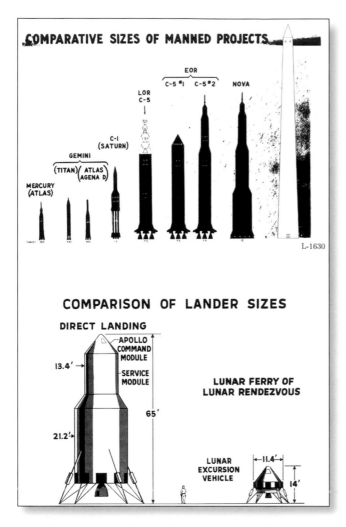

COMPARATIVE SIZES OF MANNED PROJECTS

MERCURY (ATLAS)

GEMINI (TITAN) (ATLAS AGENA D)

C-1 (SATURN)

LOR C-5

EOR C-5 #1   C-5 #2

NOVA

L-1630

COMPARISON OF LANDER SIZES

DIRECT LANDING

APOLLO COMMAND MODULE

13.4'

SERVICE MODULE

LUNAR FERRY OF LUNAR RENDEZVOUS

65'

21.2'

LUNAR EXCURSION VEHICLE

11.4'

14'

ABOVE: The early Apollo lunar lander concept for the Direct Ascent lander, seen to the lower left. It would have been very difficult to land with the astronauts reclining on their backs more than 50 feet (15 m) above the Moon and unable to see the surface. Contrast with the smaller "Lunar Excursion Vehicle" at lower right from the later Lunar Orbit Rendezvous (LOR) plan.

OPPOSITE: Construction of NASA's Vehicle Assembly Building at the Kennedy Space Center in 1965.

as a single craft that would journey to the Moon—whether as a large spacecraft assembled in Earth orbit or launched in one piece from the ground—was the simplest and most direct path from A to B that could be conceived.

A major drawback of both systems was the lander itself. The Apollo capsule would ride atop a propulsion stage with landing legs—it would be quite tall, perhaps 60 feet (18 m)—that would land tail first on the Moon. This would require the astronauts to somehow look over their shoulders, possibly using mirrors or TV cameras. Alternative designs suggested a second seating position and viewport from which a single pilot could accomplish a landing. Many designs were considered, but none were entirely satisfactory.

Add to this the monstrous requirements of building the immense Nova rocket for the Direct Ascent method, or the infrastructure required both on the ground and in orbit to launch and assemble the pieces needed for the EOR approach via multiple smaller Saturn I rocket launches, and you had the makings of one massive headache that bedeviled even the brightest engineers.

Now, place this in the context of 1961. At the time that Kennedy made his first Moon speech, NASA had just one fifteen-minute suborbital flight under its belt, powered by a tiny rocket called the Redstone that could not even place the Mercury capsule in orbit, and the technology needed to undertake the journey to the Moon with astronauts was still just a dream—today we would call it *vaporware*. The first Saturn I rocket had not yet flown, and other, smaller boosters were still exploding during launch attempts—lots of them. There were many people, including Webb himself, who had doubts that a lunar landing could be accomplished by the end of the 1960s (as outlined by Kennedy).

As Gene Kranz, at the time a young man working within NASA to get the Mercury program into full gear and who would later be the flight director during the landing of Apollo 11, recalled, "To those of us who had watched our rockets keel over, spin out of control, or blow up, the idea of putting a man on the Moon seemed almost too breathtakingly ambitious."[13] The late 1950s and early 1960s had not been kind to the rocketeers, with many a booster choosing terra firma as a fiery destination immediately after launch. The idea of reaching the Moon in nine years, much less landing two Americans there and returning them home, seemed too big a challenge. There were a few million steps to be taken between the here and now and the surface of the Moon.

Nonetheless, the work went ahead at an ever-increasing pace, with NASA expanding as the federal dollars rolled in. In 1960, somewhat ironically, NASA's budget was about the same percentage of the federal budget as it is today—0.5 percent. By 1961 it

had almost doubled to .9 percent, or about $744 million (about $6 billion today). In 1962, it was about 1.2 percent, and by 1966—the historical height of NASA's budget—it was almost 4.5 percent of the federal budget, or $5.2 billion—about $41 billion in today's dollars. With these budget increases came a huge expansion in facilities: a new launch complex in Florida, the construction of the Manned Spaceflight Center in Houston (now the Johnson Space Center), and the expansion of existing facilities at the Marshall Space Flight Center in Huntsville, Alabama, and many other places. There was also an enormous increase in the NASA-funded workforce, which grew to an all-time high of 420,000 direct employees and contractors by early 1966.

By the time Kennedy gave his second Moon speech at Rice University in 1962, the effects of the increased manpower and expanded facilities were apparent. Apollo was moving ahead, even as Mercury was flying and Gemini was in preparation, and a decision on how America would go about landing astronauts on the Moon had to be made. The choice was made in an unlikely fashion.

LUNAR ORBIT RENDEZVOUS: THE ELEGANT SOLUTION

As the debate between Direct Ascent and Earth Orbit Rendezvous continued, a third idea emerged, struggling to find its way through an increasingly bureaucratized NASA. It is important to remember that the only part of the rocket that would return to Earth was the tiny Apollo capsule. Whether just the crew-carrying capsule atop an enormous lunar lander assembled by numerous Saturn I launches for the Earth Orbit Rendezvous, or the tip-top of the lander carried by the gigantic Nova superbooster, the capsule was all that would come back to Earth. Everything else would be tossed away en route, which raised the question: Was there a more efficient way to design the disposable components so as to minimize weight and maximize capability? Scheduling would be critical as well—the program had an end-of-the-decade deadline. What was the most efficient, affordable, and speedy way to accomplish a lunar landing, now that Kennedy had put a de facto expiration date on it?

The ultimate answer lurked within the mind of an engineer named John Houbolt, who was working in comparative obscurity at NASA's facility in Langley, Virginia. Houbolt had previously been employed in rendezvous and orbital design and was thought of as "the rendezvous man" within the small orbital mechanics community. Houbolt had been burning some midnight oil, going over the various ideas about using modular spacecraft that would leave behind unneeded mass as

Early rockets exploded with some frequency. The Atlas, which is seen here detonating at launch, had a failure rate of almost 50 percent when it was being prepared to carry John Glenn to orbit.

TOP: A 1962 press conference announcing the Lunar Orbit Rendezvous decision (see page 37). From left to right, Jim Webb, NASA administrator; Robert Seamans, NASA associate administrator; Brainerd Holmes, director of the Office of Manned Spaceflight; and Joe Shea, deputy director of NASA's Office of Manned Space Flight.

ABOVE: John Houbolt discusses his Lunar Orbit Rendezvous plan in 1962.

they progressed through their flights. Similar ideas had been expressed as early as 1923, when Hermann Oberth formalized the notions first put forth by Russian engineer Yuri Kondratyuk in 1916. Then, as the US space program was just getting started in 1958, an engineer named Thomas Dolan at the astronautics division of Vought Aerospace wrote up a study called *MALLAR*, for "manned lunar landing and return." *MALLAR* included the first formal mention of a dedicated lunar lander, one separate from what became the Command Module, in the American space effort. The study had been presented to NASA but was quickly sidelined in the rush to get an American in orbit.

Houbolt unearthed the paper, studying it in detail and also reading up on Direct Ascent and EOR. Thus armed, he undertook his own detailed calculations of how a separate, dedicated lunar lander might work, an option called Lunar Orbit Rendezvous, or LOR. In simple terms, LOR would entail launching a stripped-down set of two spacecraft: an Apollo capsule with its propulsion unit and a separate lunar lander. Once in orbit around the Moon, two of the three astronauts would move from the Command Module to the lunar lander, which would eventually be called the Lunar

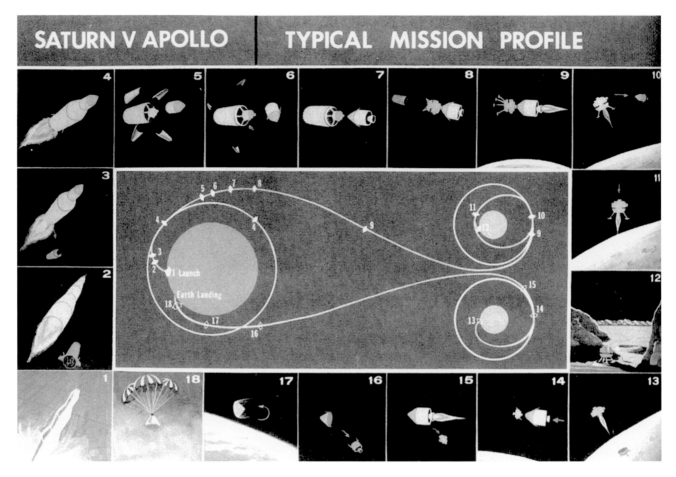

The Lunar Orbit Rendezvous plan as eventually embraced by NASA. One can see why it would appear dangerous to NASA planners in 1961.

Module, and descend to the Moon in that purpose-built spacecraft. After the lunar exploration, the upper part of the LM would boost away from the landing stage, return to orbit, and rendezvous and dock with the CM. The crew would move back into the CM, and the LM would be cast adrift as the CM returned the three astronauts to Earth.

The plan had a huge advantage in weight and fuel savings—at each stage in the journey, unneeded parts of the Apollo rocket would be left behind, decreasing the amount of mass (which would require less fuel to propel) used to complete the job. Simple.

Except that to most engineers it appeared to be *anything but* simple. In fact, to many it was terrifying. To accomplish the mission in this way required multiple rendezvous and dockings—first in Earth orbit, then twice in lunar orbit, 240,000 miles (386,000 km) away! America had barely gotten *one* man into space—and not even into orbit—and the idea of multiple spacecraft separating in lunar orbit, possibly never

to find one another again, was considered by many to be just too much to bear. Testing rendezvous and docking procedures in *any* orbit was still a few years off; this was something to be explored in the Gemini program. Committing to such an apparently risky and dangerous methodology at the time was unthinkable, and few liked the idea, including von Braun.

Houbolt was, however, undeterred. He talked to whoever would listen about his version of LOR to the point that some considered him a bit of a crank. *Let the big boys figure it out*, they thought. This was way above the pay grade of some rank-and-file engineer. Houbolt then tried to get the attention of upper management, and failed. He even bent the ear of Robert Seamans, the new NASA associate administrator, during a brief encounter in 1960. While Seamans appeared to be intrigued by the idea and eventually held meetings on the subject at which Houbolt presented his plans, little came of it. At one of these meetings, Max Faget, the feisty designer of NASA's manned spacecraft, jumped out of his seat and yelled accusingly, "His figures lie! He doesn't know what he's talking about!" Such was the reception that Houbolt received from some of his colleagues, though few were as slashing as Faget.

Meetings came and went. Committee after committee discussed the schemes for getting to the Moon—mostly Direct Ascent and EOR—and Direct Ascent was leading the pack, as it had the fewest possible downsides. The astronauts would blast off on the giant Nova rocket, sit in one spacecraft all the way to the Moon, explore the surface, then return to their massive spacecraft to leave the Moon and return to Earth. Were it not for the enormous weight penalty, and the titanic challenge of developing the Nova, this might have ended up being the course followed by Apollo, despite von Braun favoring EOR.

"DO WE WANT TO GET TO THE MOON OR NOT?"

Despite a tongue-lashing from Faget, Houbolt was confident that his calculations were correct. They showed that LOR would allow for a far smaller spacecraft, about half the mass of the other choices, that required much less fuel required to lift it. Frustrated, he once again bypassed his superiors at Langley and wrote a three-page letter directly to Seamans. It accomplished little. In November 1961, a decision on the final designs for Apollo was becoming critical now that Kennedy had announced that the goal needed to be reached by the end of 1969. Knowing this, Houbolt fired off another letter to Seamans, this one a full nine pages in length, that began with:

Dear Dr. Seamans,

Somewhat as a voice in the wilderness, I would like to pass on a few thoughts on matters that have been of deep concern to me over recent months. This concern may be phrased in terms of two questions: (1) If you were told that we can put men on the Moon with safe return with a single C-3 [Saturn rocket], its equivalent or something less, would you judge this

statement with the same critical skepticism that others have? (2) Is the establishment of a sound booster program really so difficult? . . . It is conceivable that after reading this you may feel that you are dealing with a crank.

He punctuated the letter in page three by saying, in underlined text, "Do we want to get to the Moon or not?," adding that while he realized his ideas did not necessarily conform to established thinking or "ground rules," these rules were largely arbitrary and due to fear within the ranks. Nobody wanted to seriously consider innovative alternative solutions in his view. Houbolt went on to discuss LOR in some detail, effectively lecturing the assistant administrator of NASA.

These were fighting words. This was simply not how such communications were undertaken within NASA, especially when writing up NASA's management food chain and bypassing several managerial levels.

To his credit, Seamans did not dismiss LOR, even though it had previously failed to garner much traction. He sent the proposal out for comment, with it ultimately ending up in the hands of von Braun. Despite his earlier enthusiasm of the EOR concept—and in spite of his misgiving about attempting what were perceived as dangerous rendezvous and docking operations way out in lunar orbit where anything could happen—von Braun reluctantly accepted the logic of Houbolt's conclusions, announcing them in a meeting of senior NASA leaders in spring of 1962. He summarized by saying that, in effect, the astronauts would be just as dead if something went wrong in lunar orbit as if something went wrong in Earth orbit, and that the difference in risk was, in fact, minimal. It simply *felt* more dangerous.

And that was that. With von Braun converted, LOR's momentum won the day, and despite lingering reservations, NASA selected LOR for Apollo. It was increasingly clear by then that Direct Ascent and even EOR would probably miss the end-of-the-decade deadline, and that LOR was the only way to get the job done.

Houbolt was vindicated, and Apollo had its basic mission design. Now NASA just had to build it.

CHAPTER 5

THE REAL STUFF

"WE ARE GLIDING ACROSS THE WORLD IN TOTAL SILENCE, WITH ABSOLUTE SMOOTHNESS; A MOTION OF STATELY GRACE WHICH MAKES ME FEEL GODLIKE AS I STAND ERECT IN MY SIDEWAYS CHARIOT, CRUISING THE NIGHT SKY."

—Mike Collins recalling his spacewalk
during Gemini 10

AS NASA SLOWLY DEVELOPED ITS newly determined lunar mission architecture, the cadre of astronauts who would fly to the Moon—two members of the original seven Mercury astronauts, as well as members of the New Nine and the groups that followed them—went through their paces. As Apollo progressed, the technologies and abilities that would be required to fulfill the lunar mission goals were tested in Earth orbit. This was the proving ground for the Apollo mission, and Gemini was the program that tested many of them, including the three astronauts bound to fly aboard Apollo 11.

Where Mercury had been a single-seat spacecraft designed to get an American into space and test one's ability to survive there as quickly as possible, Gemini was an entirely new, two-place capsule with startling and innovative capabilities. It had twin hatches, one for each astronaut, that could be opened in space and used to depart and reenter the

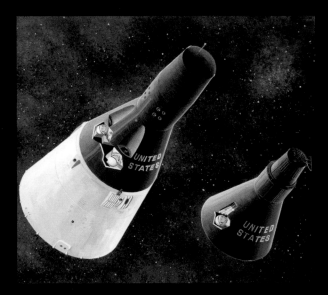

OPPOSITE: Ed White performs America's first spacewalk during the Gemini 4 mission on June 3, 1965.

ABOVE: A 1963 comparison of Mercury and Gemini spacecraft.

Gemini 7 as seen from Gemini 6, during the first-ever close orbital rendezvous, which occurred in 1965. The ability to rendezvous in space was critical to the success of Apollo.

spacecraft. It also had far better navigation capabilities and a vastly improved ability to utilize them. While the Mercury capsule could change its orientation (the direction it pointed in orbit), its only other navigational capability was to fire retrorockets, which would cause it to deorbit and splash down in the ocean at the end of its mission. The Gemini spacecraft had a full set of maneuvering thrusters, with the ability to change not just the direction it was pointing but also its orbital trajectory—it could change speed and direction in orbit, somewhat akin to a fighter jet in space. For that reason, the Gemini spacecraft was very popular with the astronauts who flew it.

The Gemini missions would test everything about flying to the Moon except landing there. This included:

- two astronauts flying together
- longer missions, simulating the duration of extended lunar missions
- life support technologies and the effects extended weightlessness

- extra vehicular activity (EVA, the technical name for a spacewalk) and performing useful tasks in space
- changing orbits, then returning to the previous orbit if desired
- changing trajectory within an orbital path to rendezvous with another spacecraft—critical for the rendezvous and docking maneuvers between the Apollo CM and LM
- docking two spacecraft together—the final critical step in rendezvous

For these and other reasons, success was critical, and each Gemini flight had a densely packed schedule of objectives. Not all succeeded, and some elements, such as successful work performed during an EVA, were elusive and met only during the final flights of the Gemini program.

The first of the Apollo 11 astronauts to fly was Neil Armstrong on the Gemini 8 mission. This had been preceded by Geminis 3, 4, 5, 6A, and 7 and was followed by 9, 10, 11, and 12 (Gemini 1 and 2 were unmanned).

Some notable accomplishments of the Gemini program were:

- Gemini 3, the first flight of a two-person spacecraft in the US program and first change of a spacecraft's orbital path
- Gemini 4, with Ed White performing the first spacewalk taken by an American
- Gemini 6A and Gemini 7, the first rendezvous performed between two spacecraft in orbit
- Gemini 8, the first docking of two spacecraft, the Gemini capsule and an Agena stage
- Gemini 11, the highest orbit attained (with a boost from a docked engine stage)
- Gemini 12, the first practical work accomplished by an astronaut during an EVA

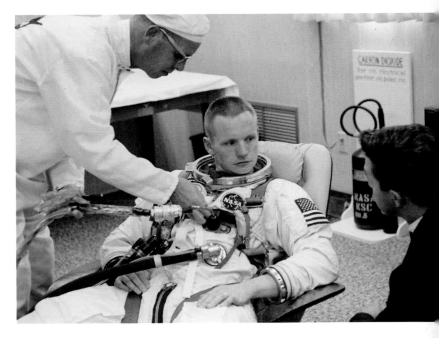

Neil Armstrong prepares for the flight of Gemini 8 in March, 1966.

It is worth mentioning that the Soviet Union had performed the first two-cosmonaut flight in 1962 and the first spacewalk in 1965. But by 1965, the US program was catching up with, and overtaking, Soviet efforts. The path to Apollo was being pioneered by the aggressive pace of the Gemini flights.

The astronauts who would fly on Apollo 11 each had their first experiences in space during Gemini. Neil Armstrong flew on Gemini 8, Mike Collins on Gemini 10, and Buzz Aldrin on Gemini 12. Each of these missions had its high and low points, and each prepared the participants for the upcoming Apollo flights in different and unique ways.

and to dock with an unmanned Agena target vehicle. But when the Agena failed to launch properly, the plan was changed, and Gemini 6 was postponed for a few weeks. This was the space race, and in those rough-and-ready days, NASA management was able to make changes more quickly and take more risks than we see today. The next Gemini mission, Gemini 7, launched first, in December, followed by Gemini 6 (which was redesignated Gemini 6A). The two spacecraft rendezvoused in orbit, coming to within a foot of each other, but did not dock as they were not equipped to do so. That left the actual docking experiment to be done on Gemini 8.

COURTING DEATH: GEMINI 8

Armstrong would command Gemini 8 with Dave Scott as pilot (the crew position descriptions can be misleading, since in NASA missions the commander usually flies the spacecraft while the pilot navigates and serves as copilot). As the sixth Gemini mission with a crew, Gemini 8 had new territory ahead—the first planned docking with another spacecraft.

This activity had been planned for Gemini 6, which had been scheduled to launch in October 1965

THE PROBLEMATIC AGENA

The Agena was a type of upper-stage rocket originally designed for use in placing US satellites in orbits after the main booster rocket had shut down. For the Gemini program it was adapted with a docking collar that would allow the Gemini capsule to link up with the Agena for rendezvous and docking practice and could be used to orient the Gemini capsule (thereby conserving the capsule's fuel supply) and for EVA practice. The Agena also had a restartable onboard rocket engine to propel the

45D

docked Gemini capsule to higher orbits. The Agena was 26 feet (8 m) long (plus the length of the added docking adapter), with a diameter of about 5 feet (1.5 m), and it had a relatively small rocket engine with about 16,000 pounds (71 kN) of thrust.

While successful as used by the air force for satellite launches, the Gemini-Agena Target Vehicle (GATV), the version modified for NASA's use during Gemini, was troublesome. Out of seven attempts, only two were fully successful, so it was not surprising that Armstrong and Scott suspected the Agena when their Gemini flight began to sour just hours into the mission.

It started out promisingly enough. On March 16, 1966, Armstrong and Scott waited to launch in their Gemini capsule at Cape Canaveral, atop an air force Titan missile, a repurposed ICBM that stood in for the still-developing Saturn IB booster. The Agena lifted off from a nearby pad at 10:00 a.m. Eastern Time on another Titan, and made orbit without complication, properly orienting itself for the upcoming docking. Then, at 11:41, the Gemini 8 spacecraft departed from Launch Complex 39 without issue. The Agena was behaving properly, awaiting the arrival of the Gemini crew. Armstrong and Scott flew a perfect trajectory to chase down and dock with the Agena—their primary objective for this mission. "We went precisely on time with our Titan," recalled Armstrong, "which was a good sign . . . because it meant that our rendezvous schedule was going to be just like we practiced for."[14] It was important that both launches occur within a few hours of each other.

OPPOSITE: An early version the Agena, the Agena A, being hoisted atop an Atlas rocket in 1960.

ABOVE: Neil Armstrong, foreground, and Dave Scott head to their Gemini 8 capsule prior to launch on March 16, 1966.

Once in orbit, the two astronauts checked all the systems and aligned the capsule's inertial guidance platform, which would instruct the onboard computer to guide the spacecraft to the Agena. The guidance computer was a miracle of miniaturization for its day,

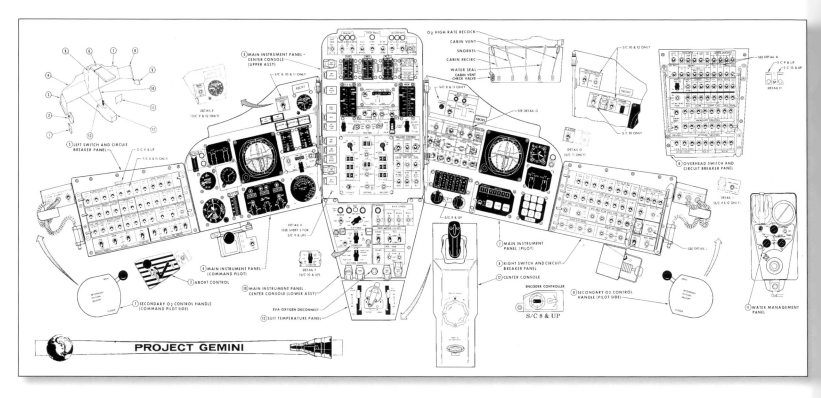

The Gemini control panel. Armstrong sat on the left side, where he rapidly scanned the instrumentation, searching for an answer to a life-threatening spin.

about the size of a toaster oven, with magnetic tape for data storage (it served as a very slow hard drive) and about enough computing power to run that same modern toaster. But it was enough, with the clever programming techniques of the time, and when mated to onboard gyroscopes and radar, it provided all the help the crew needed to find and close in on the Agena, which was just over 1,200 miles (1,900 km) away.

A few hours later they were closing on their target and took a break to eat and rest. Then the final approach began—this was the hard part—as they slowly closed the remaining distance. About four hours and forty minutes into the flight, they sighted the Agena about 76 miles (122 km) away.

They lost visual contact briefly but continued moving closer, and the astronauts could soon see the Agena's beacon light flashing against the horizon.

Armstrong and Scott were becoming more excited as they closed the distance between the two spacecraft, as reflected in the radio chatter heard at Mission Control.

Scott: "You're 900 feet [274 m] . . . 5 feet [1.5 m] per second."

Armstrong: "That's just unbelievable. Unbelievable!"

Then, a few minutes later:

Armstrong: "We're station keeping on the Agena at about 150 feet [45 m]."

For the next thirty minutes they maneuvered around the Agena to do a visual inspection—everything looked good. Then they closed the final few yards between the two craft. Moving very slowly, at about 3 inches (8 cm) per second, Armstrong eased the nose of the Gemini capsule into the docking collar of the Agena, and the docking latches snapped closed, securing them together.

Armstrong: "Flight, we are docked. Yes, it's really a smoothie."

For the next twenty minutes, the crew and Mission Control verified that the Agena and the Gemini capsule were communicating properly through their electrical connections. While ground controllers analyzed readouts, Jim Lovell, another Gemini astronaut

ABOVE: Gemini 8 does a fly-around inspection of the Gemini-Agena Target Vehicle.

RIGHT: Gemini 8 docks with the GATV version of the Agena. The trouble began immediately thereafter.

who was acting that day as the CAPCOM, sent a message up that reminded everyone listening that the Agena had been a troubled beast.

CAPCOM: "If you run into trouble and the attitude control system in the Agena goes wild, just send in the command 400 to turn it off and take control with the spacecraft."

The attitude control system comprised a number of tiny rocket thrusters distributed around the Agena. Firing these thrusters in small bursts would reorient the spacecraft. The system had been in use since the Mercury days and was considered generally reliable—but the GATV version of the Agena had proved to be finicky. Command 400 was a computer code to shut down its maneuvering system in case of an emergency.

Within minutes, Gemini 8 slid into silence as it crossed from one ground tracking station to another. At the time of the Gemini missions, NASA had a handful of tracking stations scattered around the

James Lovell (who would fly on Apollo 8 and command Apollo 13) was the CAPCOM for Gemini 8. Bill Anders (who flew on Apollo 8) sits behind him. They monitor the in-flight emergency of Gemini 8's spin, unable to help.

Earth to provide relatively good coverage of orbital flight communications, but this communication network was not seamless. From ground stations to navy tracking ships at sea, each radio dish would hand off to the next, and there were gaps during which Mission Control was out of contact. Normally this was not a problem, as the Gemini spacecraft was relatively self-sufficient. But today was a different matter, and when radio contact was regained, it carried a shocking message.

Scott: [Crackling] "We have serious problems here. . . . We're . . . we're tumbling end over end up here. We're disengaged from the Agena."

Scott's voice relayed the stress he was experiencing. During the blackout, all hell had broken loose in the skies. Shortly after docking with the Agena, Scott noticed that they were in a slow roll—something was causing the combined Gemini and Agena, firmly docked together, to change direction. Armstrong tried

to compensate by firing the maneuvering thrusters, but the conjoined spacecraft continued to drift off axis.

Recalling Lovell's recommendation regarding possible malfunctions by the Agena, Scott entered command 400 into the Agena's linked computer to shut down its maneuvering system. The Gemini spacecraft had been behaving perfectly until this point in time—surely the issue was with the often-troublesome Agena. Nothing changed. Scott entered the command again and cycled power switches to make sure the system was operating properly. But the roll continued.

Operating on the assumption that the Agena was faulty, Armstrong said to Scott, "We're going to disengage and undock." Scott agreed—they both wanted to get away from the Agena stage, which was still brimming with explosive fuels, before the torquing of the two spacecraft made separation impossible.

Armstrong released the docking latches that held the two craft together and backed away—and the spin got worse. Much worse. The Gemini capsule was tumbling along multiple axes—twisting and turning in space—at an ever-increasing rate. It was now clear that being connected with the Agena had actually been slowing their rate of spin. The problem was, in fact, with the Gemini capsule. As the engineers would later determine, a thruster control had short-circuited, and one of their Orbital Attitude and Maneuvering System (OAMS) thrusters was stuck open and firing continuously. It would do so until the tanks were empty unless they came up with a fix quickly.

America was experiencing its first true in-flight emergency, and nobody was quite sure what to do. The Gemini capsule was now spinning about once per second, which was close to the point at which the astronauts would not just be disoriented, which was dangerous, but could also black out, which would be fatal. The capsule would spin until its fuel was exhausted, and since there is no resisting air in space, it would continue to spin in orbit—with two dead crewmen aboard—for years.

Armstrong held the controller joystick in a death grip, trying everything he had learned in hundreds of hours of simulations to stabilize the capsule. Nothing worked. He quickly told Scott to give it a try—Armstrong thought he might be missing something. Scott did his best but was also unable to stop the tumble.

The situation was now critical. Radio reception at Mission Control was patchy as the spacecraft spun rapidly in orbit.

Armstrong: "We're rolling up and we can't turn anything off. Continuously increasing in a left-hand roll."

CAPCOM: "Roger."

Mission Control was not sure what else to say. Then, after a couple of tense minutes—

Scott: "Okay, we're regaining control of the spacecraft slowly, in RCS direct."

EMERGENCY REENTRY

In Mission Control, Flight Director John Hodge looked at his superiors, who told him what he already knew—the mission was over. Armstrong had used the only option left open to him—he had given up trying to control the capsule with the OAMS maneuvering system, which was clearly malfunctioning, turned it off, and fired up the Reentry Control System, or RCS. This was a second, completely separate set of thrusters and small fuel tanks designed to be used only during reentry, to keep the capsule oriented properly to prevent it burning up or overshooting the splashdown zone.

The problem was, with the OAMS system malfunctioning, and the RCS system activated, they would have to reenter immediately. Once the RCS was started, they could not be certain that the thrusters would not leak fuel, and if these valves leaked too much—or if Armstrong maneuvered too much—it would jeopardize their ability to come home.

Hodge called it; the mission was over, prepare for reentry. But they would not be reentering on schedule or on target. This was an emergency culmination of the flight, so they entered an alternate landing plan into the computer, preparing themselves to reenter and splash down near Okinawa, Japan. It was not what anyone

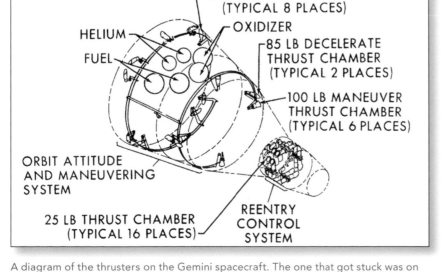

A diagram of the thrusters on the Gemini spacecraft. The one that got stuck was on the rear of the capsule.

BUZZ ALDRIN INSIDE THE LUNAR MODULE while en route to the Moon. A closer image of Buzz has been seen frequently, but this larger composite, showing much more of the LM's interior, is assembled from recently released archival images. In the right window is the 16mm movie camera that captured the dramatic film of the first decent to the lunar surface. At center top are mission checklists, and at center right and left on the control panels are the "eight ball" attitude indicators that the astronauts—pilots

all—insisted be included on Apollo spacecraft. The LM was a lightweight for its size, but it still weighed 33,500 pounds (15,200 kg) fully fueled and 9,430 pounds (4,277 kg) dry—fuel is heavy. The interior space of the small pressurized cabin was just 235 cubic feet (6.6 cu. m), of which about 160 cubic feet (4.5 cu. m) were usable—about the size of a small walk-in closet. It was a tight fit for the crew once they were fully suited up.

ABOVE: Dave Scott, left, and Neil Armstrong, right, prepare to exit the Gemini 8 capsule after their emergency return to Earth and a long wait in the Pacific Ocean.

LEFT: John Hodge was the flight controller during the Gemini 8 emergency.

Splashdown was successful, and a navy plane soon dropped divers near the capsule to attach floats that would prevent it from sinking, even if it began to take water. But that did not stop the bobbing. "The Gemini was a terrible boat," Armstrong would later say.[16]

It was a long, nauseous two hours before a destroyer showed up and the two tired, queasy astronauts, streaked with vomit and soaked in sweat, were able to clamber aboard.

Dave Scott later said of Armstrong: "The guy was brilliant. He knew the system so well. He found the solution, he activated the solution, under extreme circumstances. . . . It was my lucky day to be flying with him."[17]

Thus ended Armstrong's first mission. It would not be his last tense moment in space, however.

wanted, but there was little choice. As Armstrong said to Scott, "Okinawa . . . well, I'd like to argue with them about the going home, but I don't know how we can."[15] Scott responded in the affirmative. They both knew that this was unplanned, and the navy would have to scramble to send ships to pick them up. They could be drifting in tropical seas for some time, cooped up inside a bobbing, heaving spacecraft that would probably start heating up immediately. The Gemini capsule was a good spacecraft but not much of a seagoing vessel.

SETTING NEW RECORDS: GEMINI 10

Mike Collins had it easier. He would fly in the pilot's seat on Gemini 10 and had the good fortune of having John Young, one of NASA's most respected astronauts and a veteran of the first Gemini flight, as his commander. This was another Agena docking flight, which followed the Gemini 9 mission, and which had endured its own Agena drama. Just as Gemini 9 was supposed to launch, the Agena that was sent ahead as a docking target exploded after leaving the pad. Due to the ongoing drama with the Agena as used on Gemini missions, NASA had designed a backup, stripped-down upper stage called an Augmented Target Docking Adapter (ATDA). It was essentially an Agena without a main rocket engine, but it had small maneuvering thrusters. An ATDA was quickly prepared and launched, but once it reached orbit, it failed to release its nose cone. While Gemini 9 was able to rendezvous, it could not dock with the ATDA—the docking adapter was fouled by the snagged nose cone.

One might think that by now the Agena would be viewed as a bit of a *Flying Dutchman*, a cursed craft. But Gemini 10 launched on July 18, 1966, just an hour and a half after a flawless Agena launch, with Collins and Young aboard. Everything worked this time, with the Gemini capsule arriving in orbit about 1,100 miles (1,770 km) behind the Agena. The crew was able to rendezvous with the Agena after a bit of a struggle. Collins had some difficulty with a planned manual navigation experiment with a sextant, so they used ground-based tracking to navigate to the Agena. This consumed more fuel than was ideal, so they stayed docked longer than planned, using the Agena's fuel and thrusters to navigate. They then boosted to a higher orbit, using the Agena's main engine, and reached 412 miles (663 km) maximum altitude, the highest orbit of a manned spacecraft to date.

After a rest period, they used the Agena to position themselves in the same orbit as the derelict Agena from the Gemini 8 mission, and Collins performed the first of two EVAs. This was a simple "stand-up" EVA, with Collins opening the hatch, standing on his seat, and doing photographic experiments.

TOP: A multi-exposure photograph of the launch of Gemini 10, July 18, 1966.

BOTTOM: The Augmented Target Docking Adaptor with the fouled fairing, as seen from Gemini 9. Mission Commander Tom Stafford described it as looking like "an angry alligator."

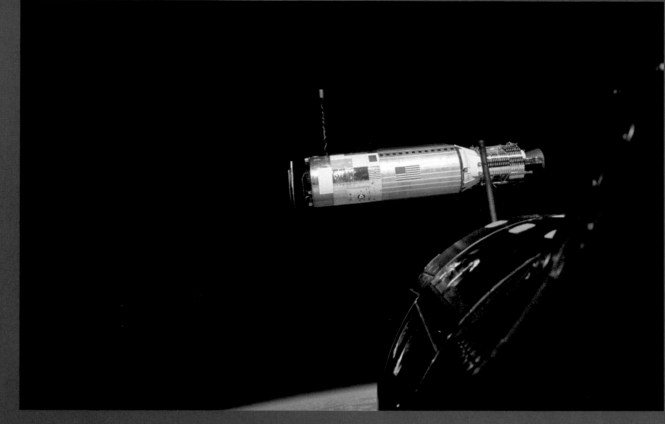

TOP: Gemini 10 closes with the first of two Agena vehicles it would dock with.

ABOVE: Mike Collins, as photographed by Mission Commander John Young, inside the Gemini 10 capsule.

RIGHT: John Young, left, and Mike Collins, right, prior to their Gemini 10 flight.

John Young, left, and Mike Collins, right, after recovery from the Gemini 10 capsule.

After another rest period, they undocked from their functioning Agena and went on to chase down the derelict one. It was almost 100 miles (160 km) away and dead as a rock—there were no navigation lights or other assistance to be had—but they ultimately found it and performed a second rendezvous. Collins then performed a much more ambitious EVA, climbing out of the Gemini capsule and floating across the gap between the spacecraft and the dead Gemini 8 Agena using a small handheld gas-thruster gun to maneuver. Collins later said that the experience was disorienting and he had had some trouble grappling with the dead Agena when he was trying to retrieve a micrometeorite experiment, but in the end it went as smoothly as something this groundbreaking could be expected to.

Seventy hours after launching, they had reentered successfully and were bobbing in the ocean awaiting pickup. The Gemini program was almost over, with just two more flights before Apollo was set to begin. Rendezvous and docking had been mastered, and the test of moving from spacecraft to spacecraft had been successful, but other aspects of EVA still eluded them. Performing useful work in space outside the spacecraft was considered essential for the Apollo program, and it was much more difficult than had been anticipated. They had only two more chances to get it right.

Gemini 11 flew in September 1966 and was successful in docking with an Agena five times. Using the Agena's rocket engine, the mission achieved an altitude record for manned orbital flight that still stands, about 850 miles (1,370 km) from Earth.

MISSION ACCOMPLISHED: GEMINI 12

As these flights were proceeding, Buzz Aldrin was preparing for his mission, Gemini 12, the last of the Gemini flights. He was determined to master the chore of performing work with tools during EVA, and he practiced the tasks tirelessly at a private high school in Maryland, where NASA had leased time in the school's pool. Again and again Aldrin descended into the deep end with partial mock-ups of Gemini and Agena hardware, staying underwater for hours in any given day, repeating the procedures over and over again. He also flew dozens of parabolic training flights in NASA's zero-g simulation airplane, affectionately dubbed the "vomit comet." The modified passenger jet would fly up at a steep angle and then nose over and dive so steeply that the astronaut inside—suited up and looking ready to go into space—was weightless for a minute or two. Aldrin practiced every procedure, no matter how minute, in both environments. Most of his fellow astronauts thought he was overzealous, and some thought his approach to EVA training a futile endeavor. But Aldrin, always one to stand by his convictions, was undeterred. He intuited that extensive practice in simulated environments was the key to EVA success, and he practiced until he could almost perform the assigned tasks in his sleep.

On November 11, 1966, Aldrin was as ready as he would ever be. He was Gemini 12's pilot, with Jim Lovell, a Gemini veteran, as the commander. At 2:08 in the afternoon Florida time, the requisite Agena target stage lifted off from Pad 14, flying into orbit perfectly, demonstrating that it had, at last, become a reliable partner for the final flights of the Gemini program.

Ninety minutes later, Gemini 12 departed the Cape. Reaching orbit a few minutes later, Lovell and Aldrin began working through the tasks required to prepare to chase down the Agena for their docking. Aldrin spent time entering data into the Gemini flight computer, a painstaking process. While the basic programs were hardwired into the system, any changeable parameters had to be entered via laborious keystrokes on a small numeric pad.

A half hour later, their radar acquired the Agena on the first try. "Houston, be advised, we have a solid lock-on, 255.5 nautical miles [473 km]," Aldrin radioed down.

Buzz Aldrin trains for the Gemini 12 mission.

Buzz Aldrin during underwater training for his Gemini 12 EVA. It was NASA's last chance to prove an astronaut could conduct work in open space before the beginning of the Apollo program.

Buzz Aldrin trained tirelessly in an underwater Gemini capsule mock-up. He is seen here practicing chores in the "busy box" at the rear of the capsule.

And then, without warning, the radar signal dropped out—it was not communicating with the flight computer. No amount of fiddling helped, and failure once again haunted Gemini's dealings with an Agena, though in this case, the problem was with Gemini's flight systems. What happened next was nothing short of amazing, and it was all done by "human wetware."

"The fallback for this situation was for the crew to consult intricate rendezvous charts—which I had helped develop—to interpret the data using the 'Mark I Cranium computer' . . . and verify all this with the spacecraft computer," Aldrin would later say.[18] In sum, he would navigate by head and by hand, using a sextant carried onboard for just such an emergency.

It had been attempted before, by none other than Aldrin's future crewmate Mike Collins, but without much success. It was fortunate for this mission and NASA that "Dr. Rendezvous," as many of the astronauts called Aldrin—some dismissively—was along for the ride.

There was a lot at stake—not only did mission success ride on this manual navigational task, but this was also something that needed to be worked out for

Apollo emergency measures in case there was a rendezvous computer failure during the lunar missions. Nobody wanted to be stuck in two spacecraft that couldn't find each other in lunar orbit.

With his charts, a sextant, and laborious calculations, Aldrin not only guided them to the Agena, but he did so using the smallest amount of maneuvering fuel to date.

They practiced docking and undocking a few times and prepared to boost to a higher orbit. But the Agena indicated possible issues with its engine, and the flight director decided to cancel the restart of the Agena's rocket engine due to the possibility of a catastrophic engine malfunction. Disappointed, Aldrin and Lovell settled in for a meal and rest period.

Twenty-one hours after launch, Aldrin performed his first spacewalk—a stand-up EVA, rising out of the Gemini hatch but not leaving the spacecraft. Like Collins had months before, he performed photographic experiments and retrieved a micrometeorite experiment from the outside. Then he had a small surprise. As he later related:

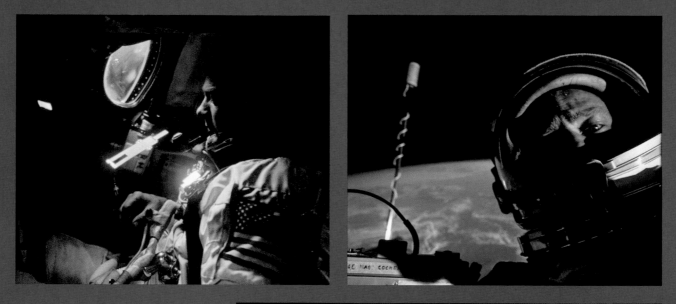

TOP LEFT: Buzz Aldrin does calculations to chase down the Agena GATV during the Gemini 12 flight. His slide rule, vital to his manual rendezvous calculations, floats in front of him.

TOP RIGHT: Aldrin during one of his two EVAs on the Gemini 12 mission.

CENTER RIGHT: Jim Lovell, Gemini 12 commander, foreground, with Aldrin, background, during the flight.

BOTTOM: Aldrin performs his second EVA for Gemini 12, working his way along the Agena.

"During the second night EVA pass I saw blue sparks jump between the fingertips of my gloves. . . . Space clearly was not an empty void. It was full of invisible energy: magnetism and silent rivers of gravity. Space had a hidden *fabric*, and the fingers of my pressure gloves were snagging the delicate threads."[19]

All too soon it was over, and Aldrin returned to the capsule for their "night" in space—their rest period before the next EVA. This was the big one. He would attempt, in this final flight of the Gemini program, to prove that an astronaut in space could accomplish routine chores. A lot was riding on this; not just the future of Apollo, but proof that Aldrin's many months of intensive and self-motivated underwater training were a valid way to approach preparing for activities in the weightlessness of space.

Aldrin, left, and Lovell, right, after the splashdown of Gemini 12. Mission accomplished.

The next "day" Aldrin hooked up to an extra-long umbilical and exited the depressurized Gemini capsule, with Lovell feeding out the hose as Aldrin moved away from the spacecraft. Carefully, Aldrin made his way across the nose of the capsule and onto the Agena to which they were still docked. Working his way slowly forward, hand over hand, he prepared a "gravity gradient" experiment for later, went back to the capsule to exchange cameras with Lovell, then moved to the back end of the Gemini spacecraft, which had a hollow adapter ring that the EVA task experiments attached to. Once he arrived, he slipped his boots into newly designed foot restraints that would help keep him properly oriented in the weightless, frictionless environment. Using the experience gained in his underwater simulations and sessions in the "vomit comet," he executed each twist of his hips and grasp of his hands with careful precision, minimizing motion and torque.

He began his assigned tasks on what had been dubbed "the busy box," an assortment of screws to be tightened, bolts to be cut, and other chores that the astronauts called "chimpanzee work"—but of course, no chimp ever had to do these in zero g. For two hours he kept at it, checking off each step in his mind as he went through the long list of tasks. Then, having displayed little exertion, he moved back to the hatch of the capsule, cleaned the windows, and went inside. After a trio of frustrating efforts in previous flights, Aldrin had shown that routine tasks could be accomplished in EVA with proper preparation and training. Each mission had contributed to the database of knowledge on the rigors of EVA, but not until Gemini 12 did anyone get it completely right.

Score two for Dr. Rendezvous.

After a couple more days and one more stand-up EVA, Lovell and Aldrin fired their reentry thrusters and headed through a fiery plume to a splashdown near Bermuda in the Atlantic Ocean. Gemini 12 had accomplished most of its mission goals, and most importantly, it had proved the value of humans in space with the manual navigation to the Agena and the first flawless EVA performance of the program.

As Aldrin later reflected, "Project Gemini had finally triumphed. *All* of its objectives had now been met. We were ready to move on to Project Apollo and the conquest of the Moon."[20]

APOLLO'S CHILDREN

"I didn't understand all the attention, either good or bad, that I was getting—I didn't think I had done anything to deserve it."

—Rick Armstrong, son of Neil Armstrong

THE SUBURBS AROUND HOUSTON, TEXAS, AND THE KENNEDY SPACE CENTER IN Florida were brand-new, tight-knit communities, where many astronauts and their families rubbed elbows with thousands of family members of others working on the space program. While kids were kids—playing football, delivering newspapers, and going to school—being the child of an astronaut was somewhat different, though many of the astronauts' children did not really realize it at the time.

Each of the three Apollo 11 astronauts had three children. Armstrong had two sons and a daughter (sadly, the daughter died as a child), Aldrin had two sons and a daughter, and Collins had two daughters and a son. Each of these children responded differently to life in the limelight as the children of the first crew to participate in a Moon landing. Here are some memories two "children of Apollo" shared in interviews with the author.

Typical suburban sprawl outside the Johnson Space Center, near Houston, Texas. It was in neighborhoods like these that the astronauts and their families lived.

ANDREW ALDRIN

Andrew Aldrin presents at a conference.

Andrew "Andy" Aldrin was born in the same year as NASA, 1958. His career path has ranged from executive positions at Boeing and United Launch Alliance to serving as president of Moon Express, a small startup in Northern California that aims to land the first private robots on the Moon. He ran the Buzz Aldrin Space Institute, which was founded to advance space exploration and development, and is currently an associate professor at the Florida Institute of Technology. While his life has taken him into aerospace as a profession, as a boy he was more interested in football than spaceflight. He describes a sense of normalcy about his childhood:

I grew up in Nassau Bay, [Texas,] which was somewhat of an astronaut enclave, so it was normal to have your dad going to the Moon or into space. My elementary school was filled with astronauts' kids. Dad was busy, and he was gone a lot, but that was his job. Where we lived, the vast majority of the people worked at NASA, and at that time, working at NASA was not a forty-hour-per-week job. So I didn't really grow up wishing my dad was home more often like everybody else's dad, because everybody else's dad was working the same kind of hours.

He remembers his father preparing for Gemini 12:

I was seven or eight when my dad flew on Gemini 12. I knew that they had experienced problems with the EVAs on the Gemini missions a few months before my dad's mission, but I didn't really understand how hard he was working to close that gap on Gemini 12. I was, however, very aware of the spacewalks and the work he had done with underwater training to prepare. Heck, when I was six, dad threw me in the pool with a scuba tank for the first time. It wasn't anything relevant to EVA—I think he just wanted to share his fascination with underwater training!

Andy was eleven years old when Apollo 11 launched. Regarding the danger of the mission, he says, "Aside from the fact that I got to crash the lunar simulator over at NASA, I really didn't have a good idea of the risks involved in that first landing. What did worry me was the ascent engine on the LM. . . . But as for the rest of it, I assumed that it would all work. Now I understand just how dangerous it was, of course." His memories of the Moonwalk are more specific:

Cables, such as the one attached to the TV camera, did not lie flat once uncoiled. They were easy to snag with the boot of the clumsy EVA suits.

> My most distinct memory is of my dad hopping around on the Moon. And I knew why he was doing this . . . he was trying to understand the most efficient way to locomote while on the Moon. So he's doing this lunar bunny hop, and I'm fixated on this cable, stretched between the LM and one of the experiments or the TV camera, and I'm just sure my dad's going to trip over that cable and end up flat on his back. I wasn't really worried about him, you understand—I knew that NASA's technology was top-drawer. But if he tripped over that cable in front of hundreds of millions of people watching on TV, and most importantly, my 200 classmates at school, that would be terrible. I was scared to death not that he might die, but that he might embarrass me. That's what goes through an eleven-year-old's mind when your dad is walking on the Moon!

Andy closed his interview with a comment about his father's ongoing passion for space-flight: "It's absolutely his life's work, and nothing is more important. He never stops thinking about it. For my dad, it's a personal passion."

ERIC ALAN "RICK" ARMSTRONG

Rick Armstrong may have spent his childhood in the shadow of the Moon, but it was Earth's oceans that called to him professionally. After majoring in biology at Wittenberg University, he worked as a marine-mammal trainer for a few years and then went to work for the US Navy in Hawaii. For the past couple of decades, his professional life has revolved around software and database design. He remembers his father, Neil, as similarly academic, yet with a sense of humility that has become legendary: "Dad read extensively and was very knowledgeable in a great many subjects, yet he wasn't afraid to admit that he didn't know much about whatever the topic of the conversation was, if that was the case." He says his father described himself as a "white-socks, pocket-protector, nerdy engineer."

While Rick remembers some of the Gemini 8 flight, he knows that it was the X-15 that really got his father engaged in going out beyond the blue sky. "Dad held his time in the X-15 program in very high regard," he recalls. The future astronaut worked closely with the engineering team on development of the X-15 flight systems and made seven flights in the rocket plane from December 1960 through July 1962. "He thought of these as team accomplishments rather than personal achievements," Rick says.

Rick Armstrong poses with a painting of his father at Edwards Air Force Base in 2014.

Rick's experience growing up was similar to Andy Aldrin's. "I never thought our family life was different than other families'. His job was always as a pilot or astronaut," he recalls. "It was a normal suburban life. . . . It wasn't until later, when I was in college, that I realized what a profound effect my father had on society. It was a bit of a shock, and rather humbling."

The perilous Gemini 8 mission was just part of his father's work, and Neil rarely spoke of it. "I'd ask Dad about work, and he didn't say much. . . . He did mention that the Gemini mission had been a challenge and required himself and Dave Scott to do some intense flying. But in general, that was about as dramatic as he would get when discussing such things." That was as humble a response to the near-disaster of Gemini 8 as can be imagined. "I did not understand at the time how much trouble Gemini 8 was in. [It] wasn't really until years later that I understood how bad it was, and how close they came to dying."

Neil Armstrong's advice to his son? "Work hard, do your very best, and respect your neighbors." It would take these characteristics, and much more, for Neil Armstrong and Buzz Aldrin to pull off the first lunar landing. Rick Armstrong's memories of that time revolve more around school. "I don't think school in Houston was much different, at least I didn't notice it. It wasn't until we moved away from Houston [after the mission] that it changed. . . . I didn't understand all the attention, either good or bad, that I was getting—I didn't think I had done anything to deserve it." As he recalls:

> I was old enough to know what was happening but not old enough to understand the complexity and risk. It was just what they [the astronauts] did; it was their job. We knew it was dangerous, but lots of kids in that neighborhood had dads with dangerous jobs. I never had any doubts or fears because I was sure everything was going to work, or if it didn't, they would figure something out to fix it.

For Rick, watching his father walk on the Moon was no big deal—perhaps because of his father's humility. "I really don't remember specific thoughts during the Moonwalk," he says. "It was just cool watching it. I knew it was my dad up there, and he seemed to be doing fine. I think we had every confidence that he would come home safe."

CHAPTER 6

MOON MACHINES I: THE SATURN V

"IT'S ALMOST AS IF JFK REACHED OUT INTO THE TWENTY-FIRST CENTURY WHERE WE ARE TODAY, GRABBED HOLD OF A DECADE OF TIME, SLIPPED IT NEATLY INTO THE '60S AND '70S AND CALLED IT APOLLO."

—Eugene Cernan, commander of Apollo 17

WITH THE PATH TO THE MOON LAID OUT by the adoption of the Lunar Orbit Rendezvous decision, and the techniques for handling specific requirements worked out in the Gemini program, there was still one enormous hurdle to accomplishing the goal of a lunar landing in the 1960s: building the machines to do it. The rockets and spacecraft of the Mercury and Gemini programs—while functional for pursuing goals such as long-duration spaceflight, rendezvous, and docking—were designed specifically to operate in low Earth orbit only. An entirely new generation of machinery would be required for the lunar effort, and the design and testing of these machines were gargantuan tasks.

Much has been written about the Saturn V Moon rocket, the Apollo spacecraft, and the Lunar Module that would land the astronauts on the Moon. They have been detailed endlessly in NASA publications,

TV documentaries, and, of course, books like this one. Nonetheless, some facts and figures bear repeating, for they illustrate the scope of the undertaking as few other descriptions can.

Let's start with the Saturn V. While smaller than the Nova rocket planned for the Direct Ascent mission design of 1960, it was still a behemoth of a machine. It was roughly the size and weight of a World War II destroyer, yet it was designed to fly into orbit and beyond.

At 363 feet (110 m) tall with the Apollo spacecraft assembly mounted on top, it was about 60 feet (18 m) taller than the Statue of Liberty. It also outweighed that monument many times over—the Statue of Liberty weighs 450,000 pounds (204,000 kg);

OPPOSITE: The first launch of the Saturn V. Apollo 4 stands poised to depart the morning of November 9, 1967.

63

the fueled Saturn V weighed 6.2 million pounds (2.8 million kg)—and it was designed to *fly.*

Of that incredible mass, about 66,000 pounds (30,000 kg) consisted of the fully fueled and provisioned Command and Service Module, and the Lunar Module was about 33,500 pounds (15,200 kg) fueled and ready to go. And, out of all that, only the CM—the Apollo capsule—would come home weighing about 12,600 pounds (5,700 kg). So just under one five-hundredth of the mass of the launched rocket would return home. Of that, the astronauts would account for about 480 pounds (218 kg), along with less than 50 pounds (23 kg) of priceless Moon rocks and soil.

The Saturn V produced 7.5 million pounds (3.4 million kg) of thrust from its five massive F-1 rocket engines, only about 1.3 million pounds (600,000 kg) more than its weight. But that weight changed rapidly—as the rocket ascended, the 521,400 gallons (2 million L) of fuel in the first stage would be long gone before it reached orbit, along with the 86,000 gallons (326,000 L) in the second stage, called the S-II. And, of course, the first stage falling away once its fuel was expended

removed much of the mass of the rocket. A third stage, named the *S-IV* due to design considerations for the original rocket, then took over to complete the trip to orbit and send Apollo on its journey to the Moon.

The flight profile looked like this:

- 00 minutes, 00 seconds: Liftoff.
- 02 minutes, 42 seconds: First-stage burn completed, first stage dropped.
- 03 minutes, 12 seconds: Second stage (S-II) ignites.
- 09 minutes, 09 seconds: Second stage is empty and drops away.
- 09 minutes, 19 seconds: Third stage (S-IV) ignites.
- 11 minutes, 39 seconds: Third stage shuts down.

About two and a half hours later, if all went according to plan, the third stage's single rocket engine would ignite again, breaking the Apollo spacecraft and Lunar Module free of Earth orbit, sending them on a trajectory bound for the Moon.

A period NASA illustration comparing the Saturn V to the Statue of Liberty. The rocket was not only much taller but also *thirteen times* heavier than Lady Liberty.

An F-1 engine test at the Marshall Space Flight Center. Five F-1 engines would ultimately power the Moon rocket.

A GIGANTIC LEAP

Listing these facts and figures on paper is one thing; designing them from whole cloth was another. When the decision to land Americans on the Moon was made in 1961, the newest and largest rocket in the US inventory was the Saturn I. It was just a faint shadow of what would be needed for the lunar flight. You could trace much of its heritage back to the V-2 rocket—the first stage of the Saturn I was a cluster of eight evolved versions of the Redstone rocket fuselage that had hurled Alan Shepard into a suborbital flight in 1961, and that Redstone was itself a close relative of an improved V-2. Some observers snarkily referred to the Saturn I as "Cluster's Last Stand." While effective, the eight fuel tanks in the first stage were heavy and less powerful than one of the Saturn V's first-stage engines, creating about 1.3 million pounds (600,000 kg) of thrust versus the the F-1 engine's 1.5 million pounds (6,770 kN). Additionally, it was only capable of sending about 5,000 pounds (2,200 kg) of payload to the Moon—less than half the mass of the Apollo capsule by itself. Clearly something larger was needed, and quickly. A giant leap in rocket design would be required.

Wernher von Braun and his team of German rocketeers were now at the helm of civilian rocketry in the US, and they responded to the challenge rapidly, devouring massive funds in the process. Something close to $7 billion of the Apollo program's $24 billion in expenditures (in 1960s dollars) went into the Saturn V, but without it, there would likely have been

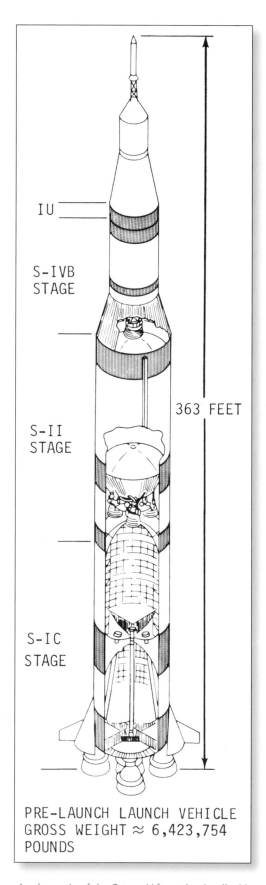

IU

S-IVB STAGE

363 FEET

S-II STAGE

S-IC STAGE

PRE-LAUNCH LAUNCH VEHICLE GROSS WEIGHT ≈ 6,423,754 POUNDS

A schematic of the Saturn V from the *Apollo 14 Flight Manual.*

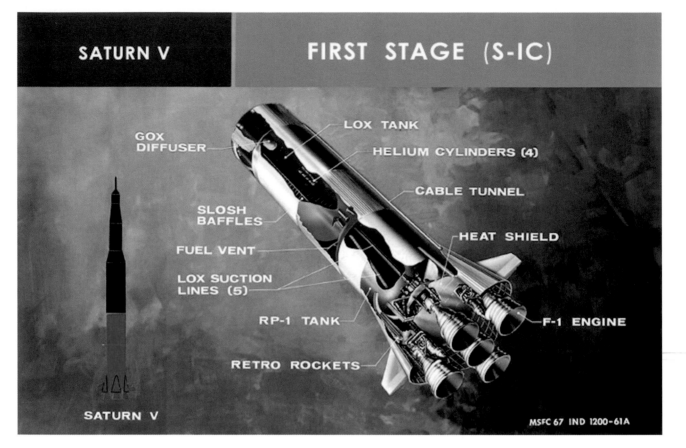

SATURN V

FIRST STAGE (S-IC)

GOX DIFFUSER

LOX TANK

HELIUM CYLINDERS (4)

CABLE TUNNEL

SLOSH BAFFLES

FUEL VENT

HEAT SHIELD

LOX SUCTION LINES (5)

RP-1 TANK

F-1 ENGINE

RETRO ROCKETS

SATURN V

MSFC 67 IND 1200-61A

A period NASA illustration of the Saturn V's first stage.

no Moon landings—the complications of launching multiple sections of the Apollo system and the crew on Saturn I rockets may well have taken longer to work out than the development of the Saturn V. Similarly, it is impossible to know if the Nova would have ever come to pass, and even if it did, it is unlikely it would have made Kennedy's deadline.

But building the larger Saturn V rocket was not simply a matter of scaling up the designs already in hand. The Saturn V was an entirely new animal. The five engines on its first stage, called the F-1, each produced the same thrust as the Saturn I's eight engines combined, 1.5 million pounds (6,770 kN). As the rocket engines got bigger, all kinds of issues popped up and had to be dealt with.

The second and third stages were smaller but used different fuels, adding another complication. The first stage used kerosene, called RP-1, and liquid oxygen, called LOX, for fuel. The second and third stages used more efficient, high-energy fuel, liquid hydrogen, and

LOX, as they were an evolution of the Saturn I's upper stage, which used those fuels. The second stage, or S-II, would have five liquid hydrogen-LOX engines, called J-2s, and the upper stage, called the S-IVB, would have one J-2. Each J-2 engine created about 232,000 pounds (1,033 kN) of thrust.

The story of the rocket engines that would power Apollo's ascent to the Moon is one of the most complex and fascinating aspects of the rocket's development. Nothing on that scale had been tried before. The first stage would employ five engines of unprecedented size, each of them over seven times as powerful as the H-1 rocket engines on the then-new Saturn I. Fortunately for NASA, the air force had paved the way for the F-1's development by contracting with rocket-engine builder Rocketdyne to develop a larger engine for its intercontinental ballistic missile program but decided not to complete the program as nuclear weapons became lighter, and no longer required huge rocket engines to loft them. NASA took over funding

FIRST ON THE MOON

the development of the mammoth engine, with a first test firing in 1959.

The F-1 rocket engine was a beast—almost 20 feet (6 m) high and just over 12 feet (4 m) across, and most of this was the rocket nozzle. Each engine weighed 18,500 pounds (8,400 kg), and when five were clustered, as they were on the Saturn V, created a total thrust of 7.5 million pounds (33,361 kN) at liftoff. They were required to burn for just under three minutes to get the Saturn V well on its way to orbit but were eventually tested for much longer durations. The F-1 was an exotic machine, employing many advances in metallurgy,

casting, welding, and more. The huge turbopumps had to move 42,500 gallons (160,880 L) of fuel per minute, which was a major technological hurdle in itself. And while those fuels were fairly well understood in application—kerosene and LOX had been used in rocket engines for years—the vast amounts being consumed in those three minutes were new territory. Add to this the hundreds of feet of plumbing required to make each engine work and that just a small metal fragment or chunk of ice could spell disaster if ingested by a high-speed turbopump, and Rocketdyne had a challenge of the highest order before them.

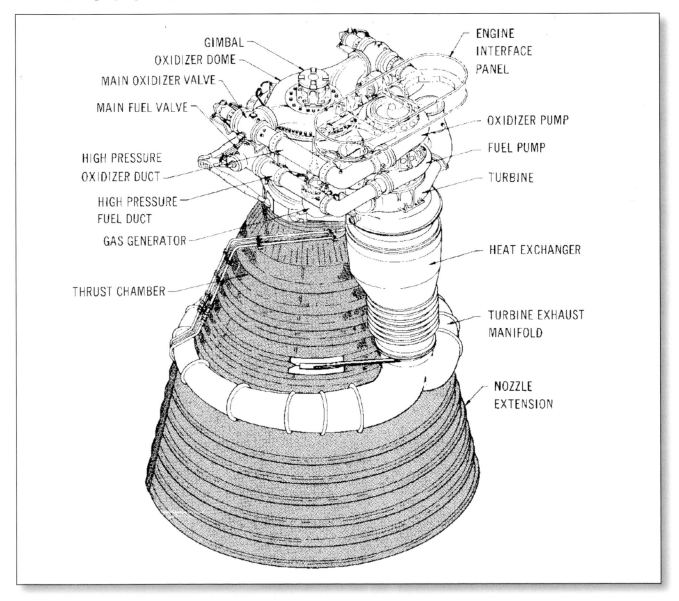

A period diagram view of the F-1 engine by NASA. It remains the most powerful single-chamber rocket engine ever built.

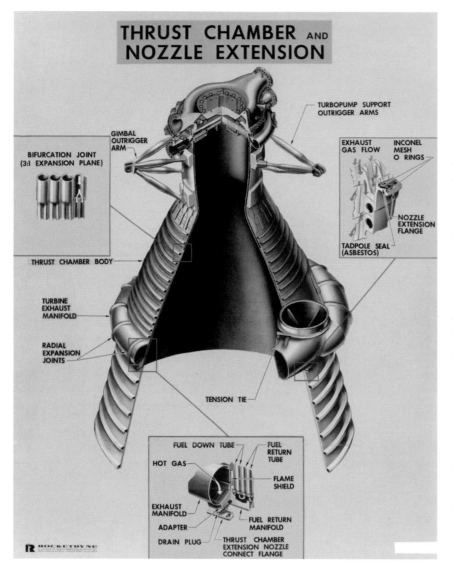

THRUST CHAMBER AND NOZZLE EXTENSION

BIFURCATION JOINT
(3:1 EXPANSION PLANE)

GIMBAL
OUTRIGGER
ARM

TURBOPUMP SUPPORT
OUTRIGGER ARMS

EXHAUST
GAS FLOW

INCONEL
MESH
O RINGS

NOZZLE
EXTENSION
FLANGE

TADPOLE SEAL
(ASBESTOS)

THRUST CHAMBER BODY

TURBINE
EXHAUST
MANIFOLD

RADIAL
EXPANSION
JOINTS

TENSION TIE

FUEL DOWN TUBE

HOT GAS

FUEL
RETURN
TUBE

FLAME
SHIELD

EXHAUST
MANIFOLD

ADAPTER

DRAIN PLUG

FUEL RETURN
MANIFOLD

THRUST CHAMBER
EXTENSION NOZZLE
CONNECT FLANGE

ROCKETDYNE

A period Rocketdyne illustration of the F-1 nozzle structure. The top half of the gigantic nozzle was made of welded, hollow tubes through which RP-1 kerosene fuel was pumped to cool the nozzle. It was a masterpiece of efficient engineering, with every component made and fitted by hand.

The engine was a study in efficiency and ingenious design. Most people who stare at the huge F-1 engines bolted in a cluster of five to the tail end of Saturn V rockets in museums located in Florida, Houston, and Huntsville don't realize that the rocket nozzles themselves are made up of coils of metal tubing. To preheat the rocket fuel, as well as to cool the nozzle, the kerosene fuel was circulated through the pipes that made up much of the rocket nozzle. It may seem counterintuitive to send explosive fluids through a red-hot rocket

nozzle, but this was an ingenious bit of efficient design. Additionally, the giant gimbals that allowed the rocket engines to swivel at the base of the Saturn V were powered by rocket fuel instead of hydraulic fluid—a dramatic weight savings— and the giant turbopump bearings were even lubricated with fuel. That's efficiency.

As previous rocket engine designs were scaled up to create the F-1, small problems became large ones, and problems that did not exist in smaller designs emerged. One of the most dramatic was called *combustion instability*— when fuel mixed in the combustion chamber (the area above the nozzle in which the fuels burned) the explosive mixture would detonate unevenly, causing early versions of the F-1 to explode during testing. In this era of slide rules and pencils, computer modeling was not yet available, and engine design was accomplished through a combination of experience, test results, and plain old intuition—a blend of art, engineering, and science. Engine after engine either shook itself to bits or exploded as the Rocketdyne engineers tried various schemes to tame the gigantic beast.

In the end, they never fully understood the combustion instability, but realized that it was causing acoustic waves to rattle back and forth within the combustion chamber thousands of times per second, ultimately tearing the engines apart during tests. The engineers experimented with changing the injector plate, a giant disk that controlled how the liquid fuels were sprayed into the combustion chamber. This disk had thousands of tiny holes drilled into it to vaporize the liquids, and these holes were moved, enlarged, and changed countless other ways in an attempt to force the

Early F-1 engine testing at Edwards Air Force Base in 1959. For the next few years, F-1 engines and turbine components either shook themselves to pieces or exploded during test runs.

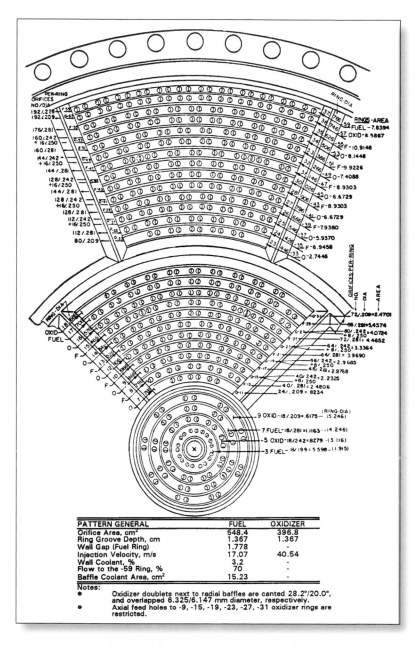

PATTERN GENERAL	FUEL	OXIDIZER
Orifice Area, cm²	548.4	396.8
Ring Groove Depth, cm	1.367	1.367
Wall Gap (Fuel Ring)	1.778	-
Injection Velocity, m/s	17.07	40.54
Wall Coolant, %	3.2	-
Flow to the -59 Ring, %	70	-
Baffle Coolant Area, cm²	15.23	-

Notes:
- Oxidizer doublets next to radial baffles are canted 28.2°/20.0°, and overlapped 6.325/6.147 mm diameter, respectively.
- Axial feed holes to -9, -15, -19, -23, -27, -31 oxidizer rings are restricted.

A schematic of the F-1 fuel injector plate. By the time the engine problems had been solved, many configurations of both the injector holes and the baffles (the injector holes are shown here drilled in sets of two along concentric arcs, and on each side to right and left is a baffle) had been attempted.

explosive fluids to burn evenly. They finally settled on a series of ribs added to the back end of the plate to help direct the spray, and this ultimately saved the day—by 1965 the F-1 was ready to go into space.

THE SECOND STAGE: UNGLAMOROUS, BUT CRITICAL

But the Saturn V was not fully compliant yet. The second stage, the S-II, was another challenge. NASA had awarded the contracts for the first and third stages—the S-I and S-IVB—prior to selecting North American Aviation in Downey, California, to build the S-II stage, so the weights and specifications of the other stages were set. Add to this that the Apollo Command and Service Module and the Lunar Module were growing in weight by the month, and mass had to be reduced somewhere in the rocket. The S-II stage bore the brunt of much of this weight-reduction program, and designing it was a nightmare.

To propel the Apollo spacecraft into orbit, the S-II needed to support and channel the power of the five smaller J-2 rocket engines attached to its base, and avoid being crumpled as this force pushed the S-II into the heavy masses above it. Were weight not such an issue, this would not be a problem—just make the hull of the S-II stage thicker and stronger. But due to the ever-more-demanding weight restrictions, the engineers had to come up with innovative new ways to design a lightweight yet incredibly strong rocket stage, and some of their solutions were truly remarkable. In effect, they had to design from the inside out, so the first task was to contain and control the vast quantity of rocket fuel needed to power the S-II's five J-2 liquid hydrogen–LOX rocket engines.

American aerospace contractors had been working with cryogenic fuels for years, but these ultra-cold liquids were still tricky to handle when the S-II program began in earnest. The art of storing liquefied hydrogen at –423 degrees Fahrenheit (–253°C) and liquid oxygen at –297 degrees (–183°C), while keeping the fuel tanks incredibly light, was in its engineering infancy. Previous efforts, such as the Centaur cryogenic rocket stage used on the Atlas rocket, utilized the same "pressure-stabilized" design as the Atlas rocket's main stage—a long, stainless steel tank that could only support its own weight when either filled with fuel or "inflated" with nitrogen gas. If it was set on end while empty, the entire structure could crumple. This would not do for the S-II stage—it needed to be far stronger to perform its task—so an entirely new design was called for.

RIGHT: The Saturn V's S-II stage being hoisted onto a test rig at the Mississippi Test Facility (now NASA's Stennis Space Center) in 1965.

BELOW: The F-1 engine injector plate. While it may not have been the ultimate cause of the combustion instability, it ended up being the solution. The baffles are visible radiating from the center.

BOTTOM: In contrast to the schematic, in practice the modifications to the injector plate were a bit more ad hoc as dozens of patterns were tried. Note the partially drilled holes to the right of center—someone changed their mind.

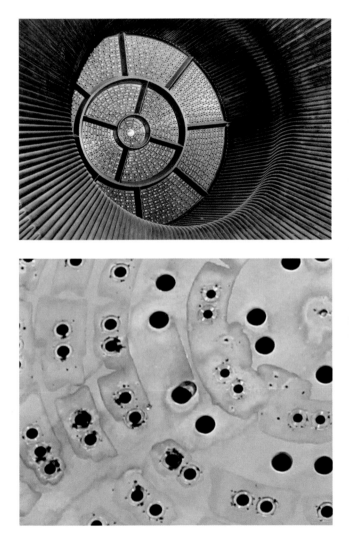

The S-II was about the size of a grain silo, standing 82 × 33 feet (25 × 10 m). The first stage of the Saturn V would push against it from the bottom with 7.5 million pounds (33,361 kN) of thrust, and atop it would ride almost 350,000 pounds (160,000 kg) of the S-IV stage and the Moon-bound Apollo machines. Given the constraints the engineers were under, it was a bit like balancing a bowling ball on top of an egg while pushing up from below—a truly daunting task.

The first order of business was to design the fuel tanks. Traditionally, the fuel and oxidizer were each contained in a separate, cylindrical tank—the kerosene or liquid hydrogen in one tank, and the LOX in the other. But this design—with two separate tanks, each with its own top and bottom—was heavy. This design could be used for the Saturn's first stage, but with the weight limitations on the S-II, this would not work; it would be too massive. So the engineers devised a

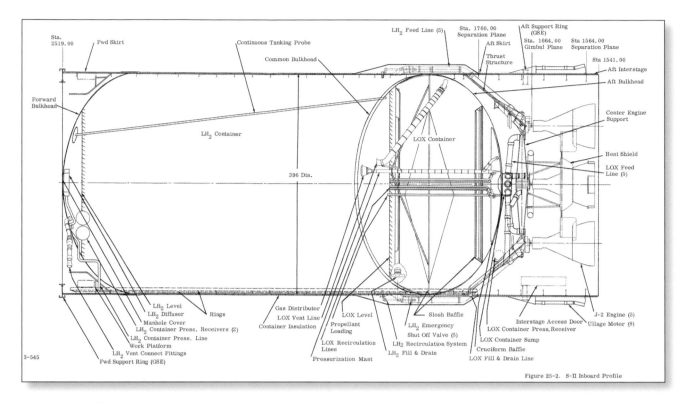

A schematic of the Saturn V's S-II stage.

common tank wall that served to separate one larger tank—essentially the entire volume of the stage—into two halves. The top two-thirds would carry the liquid hydrogen, the bottom third the LOX. This saved both weight and space by eliminating two of the tank end caps and combining them into one part, which allowed the stage to be shorter as well as lighter. And these "tanks" were no longer separate items from the fuselage of the rocket—the hull itself would be the fuel tank.

This was unknown territory for rocket designers—at 33 feet (10 m) in diameter, the S-II would be by far the largest rocket stage ever attempted that used a common wall to separate cryogenic fuels. And the fuels were stored at different temperatures, with roughly a 126-degree-Fahrenheit (52°C) differential. It would require exotic insulation technologies to prevent the "warmer" LOX from causing the far colder liquid hydrogen to boil off. If fuel loss did occur, extended storage of the fuels would be impossible.

The solution—again, never tried on this scale—was to emplace a plastic honeycomb structure between two dome-shaped disks of very thin aluminum. North American Aviation had pioneered the use of honeycomb

structures in the B-70 supersonic bomber a few years before and was confident that it would work for the Saturn V. The engineers proceeded to design and test just such a structure for the S-II.

Fabricating the domes that sandwiched that honeycombed plastic core was another challenge—they were thin and 33 feet (10 m) across. The domes were constructed by fabricating pie slice–shaped pieces of aluminum sheeting (called *gores*), each a bit longer than 16 feet (5 m), which were then welded together into two 33-foot-diameter (10-m) disks. These completed disks had to be dome-shaped to be strong enough to do their job, and to add to the complexity, the pie slices used to make the domes were not of consistent thickness. They were about 0.5 inch (13 mm) thick at the wide rim, tapering to a flimsy 0.32 inch (8 mm) at the pointed tip. Forming these gores into double-convex shapes curved from end to tip as well as side to side would be impossible using conventional, established methods. Staring at the blueprints, the engineers realized that, like so much of Apollo, there was not even a machine that could *make* the machine needed to form the gores. Each innovation seemed to cause a new engineering challenge.

The solution to this particular problem turned out to be unique, to say the least.

The company found a 60,000-gallon (227,000-L) water tank at Marine Corps Air Station El Toro, just across Los Angeles County from its headquarters, and arranged to use it to manufacture its new and weird rocket parts. At the bottom of the tank, engineers placed a mold that was the exact shape and curvature of the aluminum gores. They would then fill the tank with water and carefully lower the thin, flat pie-slice gore blanks into the water until they floated above the curved mold. They then lined the top of the metal sheets with Primacord®, an explosive rope that the military uses to breach walls and decapitate large trees. After clearing the area and a performing a

short countdown, the Primacord was detonated, and—*kapow!*—the explosion forced the metal sheet down onto the mold—almost. It took three rounds of explosives to form each set of gores for the intertank wall.

With these thin, wobbly parts properly formed, the engineers laid them out inside a sealed, dry tank and inflated them with air from below so that they would hold their shape during welding. Fabricators started welding them together into the two disks that would sandwich the plastic honeycomb. Welding aluminum is tricky at the best of times, but at this thinness and size, the first set of gores deformed where they were welded together. So it was back to the water tank, and many explosions and aluminum sheets later, a new batch of gores emerged. The engineers put them

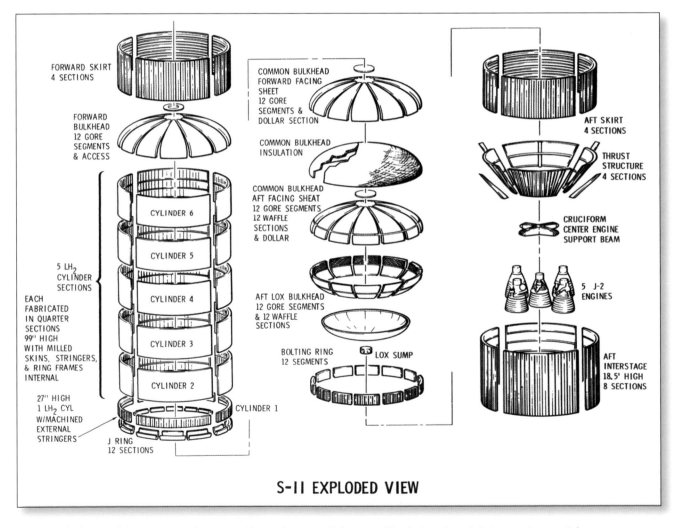

An exploded view of the Saturn V's S-II stage. The tricky "gores" that gave North American Aviation engineers nightmares can be seen at top and center.

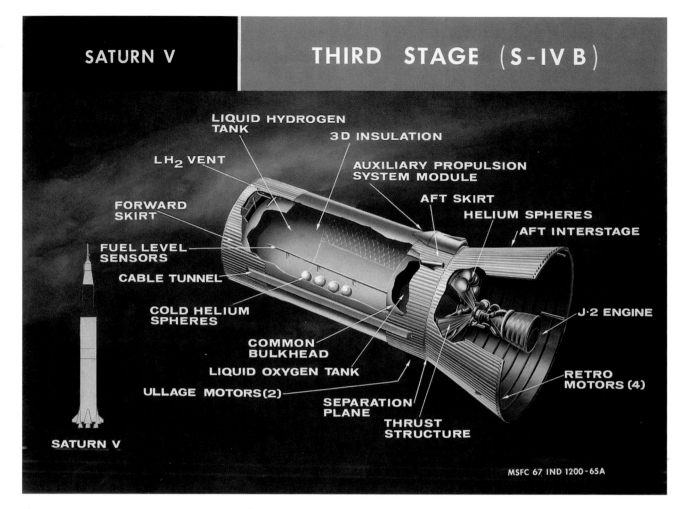

LIQUID HYDROGEN TANK

3D INSULATION

LH₂ VENT

AUXILIARY PROPULSION SYSTEM MODULE

FORWARD SKIRT

AFT SKIRT

HELIUM SPHERES

AFT INTERSTAGE

FUEL LEVEL SENSORS

CABLE TUNNEL

COLD HELIUM SPHERES

J-2 ENGINE

COMMON BULKHEAD

LIQUID OXYGEN TANK

ULLAGE MOTORS (2)

SEPARATION PLANE

RETRO MOTORS (4)

THRUST STRUCTURE

SATURN V

MSFC 67 IND 1200-65A

The Saturn V's S-IVB stage. It was actually the first stage of the Saturn V to be developed, as it was originally slated to fly atop the Saturn I.

into the dry tank, inflated them from below, and tried once more to successfully weld them together. The welds had to be perfect, and once again, conventional methods failed them.

Clever technicians devised another new machine that could slowly drive along the seams as it welded them—it was the only way they could think of to make consistent, perfect welds between the gores. After many attempts and dead ends, the technicians were successful. They wrapped the hull of the S-II stage around the completed intertank wall, adding a pressure dome at the top and another pressure dome on the bottom to which the five J-2 rocket engines were attached. This resulted in an ultralight, very strong rocket stage.

The third stage was of similar construction on a smaller scale—the Saturn V tapered to just 22 feet (6.7

m) near the top. Even though it was the third and last rocket stage, it was called the S-IVB due to a numbering scheme first devised in the Saturn I days (it was originally planned as the fourth stage of the never-built Saturn C-4 rocket). The S-IVB stage was actually the first part of the Saturn V to be contracted, with work starting on it in 1960. It used a similar common intertank wall, similar to the S-II's design, and a single J-2 engine, burning the same fuel mixture as the S-II stage. The S-IVB was contracted to the Douglas Aircraft Company (later McDonnell Douglas Aerospace) and caused similar problems but on a smaller scale. One challenge unique to this stage, though it was not specifically Douglas Aircraft's problem, was the engine, which, like the J-2s on the S-II stage, was built by Rocketdyne. This one had to be restartable in space, which was problematic. The only previous restartable engine burning these exotic fuels was the predecessor to the J-2, the RL-10. Developing the RL-10 had been an ordeal, and it only developed about 10 percent the thrust of the J-2. Further, the RL-10 only flew successfully in 1963, so the engineering behind it was still quite immature when development of the J-2 was being undertaken. This too was new territory.

There was also the problem of engine run-time. You will recall that the F-1 engines on the first stage burned for less than three minutes. The five J-2 engines on the S-II stage would burn for six minutes. The J-2 on the S-IVB stage would burn for two minutes to finish boosting the Apollo spacecraft into Earth orbit. It would then be required to burn an additional *six and a half* minutes to boost the spacecraft into a lunar trajectory, breaking free of Earth orbit. Six minutes was an eternity for large rocket engines at the time, so testing the J-2 engine for these durations and beyond was of critical importance. The first test firing was accomplished successfully in October 1960 at Rocketdyne's test facility in southern California, when it burned for four minutes. The contractor started manufacturing J-2s in 1963, and a year later, in late 1964, the engine was fired for almost seven minutes. These and other tests demonstrated the engines' ability to function long enough to perform their assigned function with reliability. By 1966, NASA was confident enough to have ordered 155 of these rocket engines.

The next step was to test and certify the J-2's ability to be restarted after sitting in orbit for three hours, the amount of time between launch and leaving for the Moon on an Apollo mission. It took a few orbits for the Apollo spacecraft to be properly aligned for trans-lunar injection (TLI), the maneuver that would break it free from Earth orbit and send it on its way to the Moon. First, there had to be tanks containing pressurized helium added—the helium would push the fuel down to the turbines and into the combustion chamber, for at this point the liquid hydrogen and LOX would be floating around in their tanks, big globs of frigid liquid that had to be forced down into the engine. To assist this further, small rockets—called *ullage motors*—were attached to the S-IVB stage to nudge it forward at a low rate of speed, propelling the rocket stage forward while

Rocketdyne's J-2 upper stage liquid hydrogen–liquid oxygen restartable rocket engine during test firing.

the liquid fuel drifted down toward the turbine intakes. After lots of testing and modeling on the primitive computers of the day, these tests were successful and the J-2 was ready to go, with the engineers confident that the rocket engine would reliably restart in space.

There were still hundreds of other systems and subsystems on the Saturn V that needed attention, and perhaps the most critical of these was the rocket's guidance computer. While the Command and Lunar Modules each had their own guidance and navigation computers, it was deemed essential that the Saturn have a separate unit of its own to assure a successful ascent into orbit, where the other computers would take over.

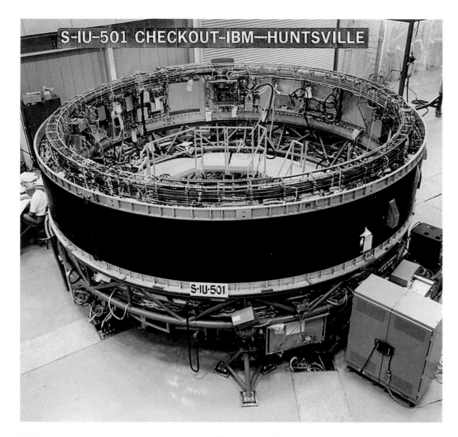

IBM's Instrument Unit, the computerized brains that flew the Saturn V into orbit.

This large guidance computer was installed in the Instrument Unit, and it was comprised of a number of metal boxes filled with electronics that lined an interstage ring that topped the S-IVB stage. It was completely separate from the computers in the CM and would be capable of flying the rocket successfully into orbit unassisted in the event that the other computers failed—as occurred briefly during the launch of Apollo 12 when the Saturn V was hit by lightning during its ascent, knocking out the CM's computer.

The Instrument Unit contained a basic digital computer, analog flight control systems, hardware to detect emergencies in the rocket's systems, and a set of gyroscopes and acceleration detectors to tell the rocket where it was, what direction it was heading, and where it should be pointed to achieve its goal. While considered redundant by some government bean counters, the Instrument Unit earned its pay on that day when the Apollo 12 capsule went dark. With the CM's electronic systems failing instantaneously, the only thing that saved the crew from a dangerous abort maneuver—which could have severely injured them—was the Saturn V's ability to fly itself until the astronauts fixed their electrical problems.

These and the myriad other systems in the Saturn V were designed, tested, refined, and tested again throughout the early 1960s, until the first unmanned test flight of the rocket occurred in November 1967. That flight, dubbed Apollo 4, replaced an incremental series of tests planned by von Braun. Traditionally each test flight, one or more of the upper stages would be filled with inert mass—usually water—to simulate fully fueled rockets, and in this way, each stage could be tested incrementally, with flaws and errors detectable stage by stage. This was the German way of doing things. But George Mueller, a senior engineer in the Apollo program at the time, argued that this would waste time and money, and that NASA should adopt a program of what he called "all-up" testing— just fuel everything and test the rocket all at once. This was, apparently, the American way of doing business. He pointed to the US Air Force's success with this technique in developing the Minuteman ICBM, and after some pushback from von Braun's cadre of engineers, Mueller got his way. At least four flights of the Saturn V were rolled into one, and Apollo 4 was a rousing success. Each stage of the rocket performed properly as it flew into orbit, and the S-IVB's engine was restarted, punching the empty Apollo capsule into an orbit more than 9,000 miles high (14,500 km). The capsule was then forced to reenter the Earth's atmosphere at speeds in excess of those it would endure during a return from the Moon, thoroughly testing the heat shield. There would be only one more test of the Saturn V before astronauts climbed aboard for a trip to orbit the Moon on the Apollo 8 mission in 1968. It was one of the most efficient and successful rocket-development programs in history.

The launch of the unmanned Apollo 4, the first test of the Saturn V rocket.

"THE ROAR IS TERRIFIC!"

CBS news anchor Walter Cronkite was at the Cape for the Apollo 4 test flight. He was an avid supporter of the lunar landing program and had been present for many rocket launches prior to this, including the smaller Saturn IB. But nothing could prepare him for the fire and fury of a Saturn V launch. As he sat in the CBS news observation center, which was possibly closer to the launch pad than it should have been, he looked on in awe, then in increasing concern, as the rocket ascended and the sonic concussions caused by its five F-1 engines slammed into the building.

"Our building is shaking here. . . . Our building's shaking!" A concerned Cronkite and a technician ran to the widow, now rattling violently in its frame and threatening to shatter from the sustained roar of the rocket. They spread their hands to support the shaking glass. Cronkite was now yelling over the thundering roar that was coming through the window. "We're holding it with our hands! Look at that rocket go into the clouds at 3,000 feet [900 m]! . . . You can see it. . . . You can see it. . . . Oh the roar is terrific!"[21]

The Saturn V was ready for its lunar mission. Now the rest of the program had to catch up.

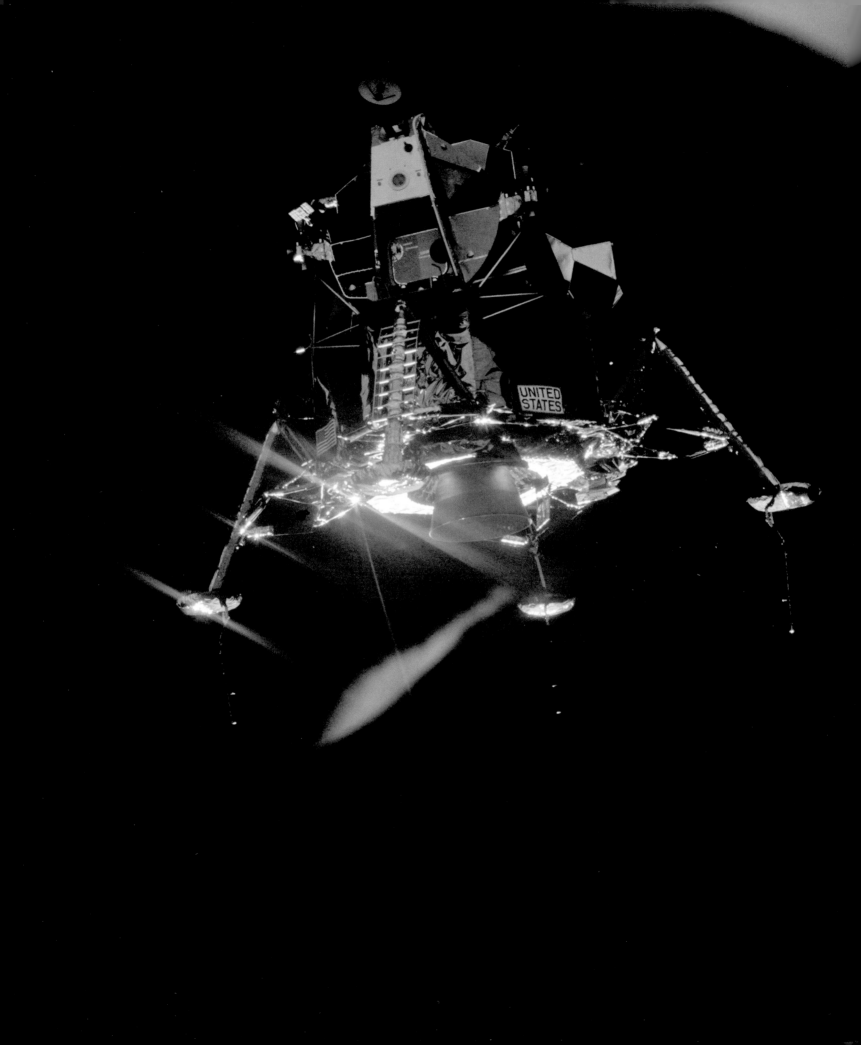

CHAPTER 7

MOON MACHINES II: THE LUNAR MODULE

"YOU'VE GOT A FINE-LOOKING FLYING MACHINE THERE, *EAGLE*, DESPITE THE FACT YOU'RE UPSIDE DOWN."

—Mike Collins, upon inspecting the LM after undocking

THE APOLLO LUNAR MODULE (LM)

was *just enough* of a spacecraft to do its job. It was just large enough, just powerful enough, and perhaps most notably, just *strong* enough to transport two humans to the lunar surface and back up into orbit. It had just enough supplies to keep them alive while they were there (with a bit to spare) and carried just enough tools and cargo to allow them to complete their tasks.

The most important consideration for the LM's designers was weight. While its job was to safely transport the crew to and from the Moon, the LM needed to be as light as possible. Every extra ounce on the LM would require more fuel to lift. Under NASA's orders, the LM's designers worked obsessively from day one to keep their Moon machine trim.

The final design was almost frighteningly fragile. While it weighed about 33,000 pounds (15,000 kg) fully fueled and ready to go, it was very lightly built for its size. The spacecraft was 23 feet (7 m) high

and at the widest spread of its landing legs 31 feet (9.5 m) wide. It had a 235-cubic-foot (6.5-cu.-m) pressurized cabin for its crew. There was a rocket motor to land it on the lunar surface and a smaller motor to get it back into orbit. The LM also had to carry enough fuel to power both those rockets, plus all the tools, supplies, and accouterments to support the two passengers on the Moon for twenty-two hours plus a bit more time before landing and after returning to orbit. (Later LMs carried more expendables for longer stays on the Moon.) A safety margin was added to this, resulting in more supplies and mass. The LM was powered by batteries, not fuel cells like the Command Module, and those were heavy as well. Radars, radios, computers, and so much more made up this revolutionary spacecraft, and these were not the compact, miniaturized units

OPPOSITE: The Apollo 11 Lunar Module *Eagle*, Lunar Module #5, preparing for its historic descent to the lunar surface on July 20, 1969.

we know today, but massive, component-and-wires, old-school electronics—the LM was designed in an era when many electronic appliances used vacuum tubes to do the work electronic chips do today. Add to this the water needed for drinking and cooling the electronics, food, still cameras, TV cameras, movie cameras, lunar surface experiment equipment, rock hammers, EVA suits, life-support backpacks, and hundreds of other items large and small, and the weight added up quickly.

The story of how this one-of-a-kind space machine was created is complex in detail, but the broad strokes are relatively simple to tell. It all started in 1962, when NASA invited US aerospace contractors to bid on the construction of its new lunar lander. Nobody knew what it should look like; few were even clear on what the specifications and limitations would be. There were many unknowns at that time; the Lunar Orbit Rendezvous decision had only been made official in July the year before. Nine companies entered the competition, and in just two months, NASA selected Grumman Aircraft Engineering. That's how quickly things were moving in the space-race era.

THE "GRUMMIES"

The "Grummies," as the employees of Grumman often referred to themselves, were elated; yet within eighteen months, many wished they had never heard of the Lunar Module. The creation of this unprecedented machine was challenging the engineers, to say the least. It was tough to design and would be tougher to build.

Other companies won contracts to build the Lunar Module subsystems. Aerospace giant TRW (Thompson Ramo Wooldridge) would manufacture the rocket engine for the descent stage, Bell Aerosystems would fabricate the engine for the ascent stage, Raytheon would build the navigation computer, Hamilton Standard would build the life-support system, and MIT's

TOP: The earliest known model of the Apollo Lunar Module, made from wood and metal paper clips.

BOTTOM: Within months of beginning work, the engineers had evolved the overall shape of the Lunar Module to the one shown here. This iteration would not last long as mass-reduction efforts continued.

A somewhat fanciful shot of the early Lunar Module design on the Moon.

computer lab would design the landing software. But Grumman would build the lander and integrate all the subsystems—the failure of any one of which could result in two dead astronauts on the Moon or in its orbit.

With the contract secured, work started immediately. The model that had helped Grumman to prepare its bid was simple—just a small wooden mock-up with a dowel sticking out the front to represent the hatch and five paper clips stuck into the sides representing landing legs. It was a humble beginning for an incredibly complex machine, and the engineers had a long way to go.

Within months the working design had evolved into a white-painted ovoid crew cabin mated to a cylindrical landing stage. It had lots of windows and five landing legs. Before long, all but two of the large

windows would vanish, as would one landing leg—the machine, as originally designed, was simply too heavy.

By the time Grumman began working on the LM, most of the other components of the program were either already contracted out or at least being designed. That meant that, in effect, Grumman had to work with whatever weight allowance was left over after all the other components were accounted for—the Saturn V was only capable of lifting a certain amount of weight. Considering that nobody had ever built anything remotely like a lunar lander before, this weight allowance was especially challenging, since the engineers working on the LM were truly venturing into unknown territory. Up until the year prior, few were even thinking about the possibility of building a dedicated lunar

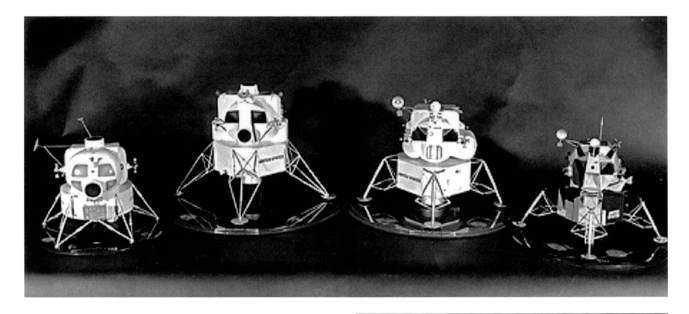

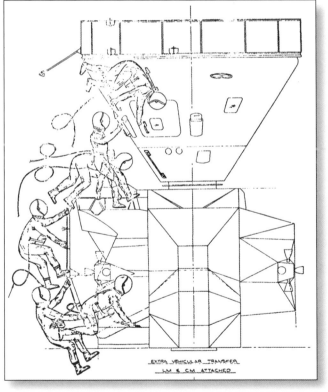

ABOVE: The evolution of the Lunar Module design can be seen in this rather hastily arranged photo. The earliest form, from 1961, is seen to the left, and the mostly final iteration, from about 1964, is seen to the right. It was a dramatic set of changes in a short span of time.

RIGHT: This illustration of a planned emergency EVA test—in which an astronaut exits the Lunar Module from a hatch, crosses through open space, and then enters the Command Module hatch—shows that the Lunar Module and Command Module were roughly the same size. Their masses and configurations were, however, vastly different.

lander—the plan had been to land a one-piece Apollo capsule on the Moon, then use it to come home. The idea of a separate, disposable landing craft was new . . . and intimidating.

THE FATHER OF THE LUNAR MODULE

Tom Kelly led the team that built the Lunar Module program at Grumman, and was every inch a visionary. Kelly was in his early thirties at the time, and with about 7,000 engineers, technicians, and craftsmen working under him, was largely responsible for the project's success. Kelly's relationship with Grumman started when he entered college in 1946, attending on a Grumman company scholarship. By 1951, he was working on missile programs at the company, and except for a call to service with the air force in the late 1950s (he was in the

ROTC while in school) and a brief stint at Lockheed, Kelly spent his career at Grumman.

He had been a believer in Lunar Orbit Rendezvous long before NASA arrived at the same conclusion, and when Grumman received the contract, the company had been working under that assumption for some time. The knowledge gained by modeling its bid around Lunar Orbit Rendezvous allowed Grumman to provide NASA

A photograph of JFK standing in front of an early mock-up of a lunar lander demonstrates just how far the design came from its earliest days.

one of the most thoroughly prepared proposals, which was reportedly instrumental in its success.[22]

"The Command Module was totally dominated by the need to reenter the Earth's atmosphere, so it had to be dense and aerodynamically streamlined," Kelly later said. In contrast, the Lunar Module had a very different job. "It [had] to be able to land on the Moon and operate in an unrestricted environment in space and on the lunar surface. [This] ultimately resulted in a spindly, gangly-looking, very lightweight vehicle that was just the opposite of all the attributes of the Command Module." He added, "There were no precedents for what we were doing, and we were neither bound nor guided by convention."[23]

One of the challenges before Kelly was Grumman itself. The company was primarily an aircraft manufacturer, known for its small passenger planes and the tough fighter aircraft it built for World War II, and thereafter jet fighters for the air force. Neil Armstrong flew a Grumman F9F in combat over Korea. But building lunar landers was not like building jets—a fighter jet contract might net thousands of orders. There would only be fifteen Lunar Modules built, so the economics of the LM were completely backward from a corporate perspective.

There was another fly in the ointment, one less quantifiable but profound nonetheless. Many of the engineers working for Kelly were career *aircraft* designers, and the LM was far different from an airplane; it would never fly in an atmosphere. More than one senior engineer looked at the blueprints and said something to the effect of, "Look at that thing! There's stuff sticking out all over! It's going to break!" Clearly, Kelly had some convincing to do, and some hard personnel decisions to make. It would take every ounce of his managerial skills, and more, to see the program through. He had to break through traditional thinking.

"In the process of simplifying the systems, we realized that we had just fallen into accepting some basic things that weren't necessary, like symmetry," Kelly said.

A mid-development iteration of the Lunar Module at Grumman's Bethpage, Long Island, plant.

"We found out that we originally had four propellant tanks in the ascent stage because it gave us a symmetrical configuration. Then we said, 'Gee, it doesn't have to be symmetrical.' . . . [It] ended up looking like it had the mumps on one side."[24]

There was another concern voiced by some of those engineers, one that rang truer than comments about the awkward appearance of the Lunar Module. With jet airplanes you had a test regimen—up to hundreds of hours of flight test, to wring out the bugs in a design. The plane had to be good but not perfect the first time; it would be refined through a proper test program. The Lunar Module, on the other hand, would only be flown for a few hours in manned test mode before the first landing. Add to this that all the major systems had to work perfectly—not 80 percent, not 90, but *perfectly*—and you can see why the program caused quite a lot of indigestion and skipped heartbeats.

PUTTING THE LM ON A DIET

Once the basic design was settled, the weight reduction program began immediately. Grumman labored to get pounds, then ounces, then milligrams shaved off the lander. The first targets were pretty obvious. As noted, the many windows on the prototype design weighed a lot—these needed to be made from multiple sheets of high-strength glass with a complex sealing mechanism, so the weight of each square inch of glass added up quickly. By the time the engineers were done, the four large windows of complex curved glass had been reduced to two small ones, with an additional tiny docking window on the roof. This saved a lot of weight, but not nearly enough.

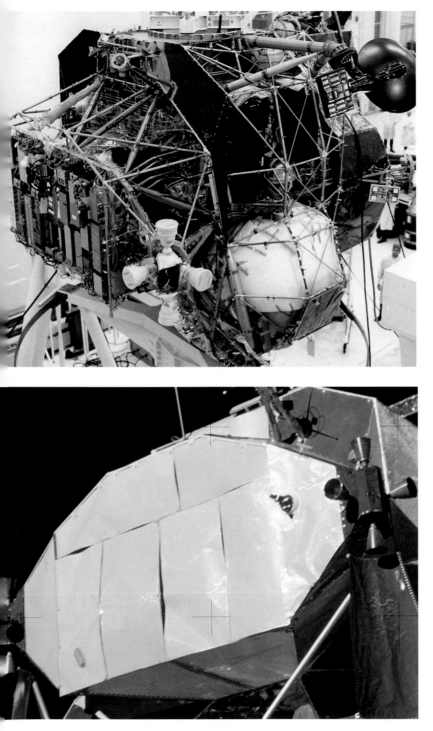

TOP: The Lunar Module was very lightly built—it was a study in form following function.
BOTTOM: The thin panels on the backside of the Lunar Module's upper stage were not structurally critical. They did, however, protect the delicate electronics and systems behind the crew cabin. This image, taken after landing, shows how thin they were.

Next to come under scrutiny were the seats. One of the engineers suggested that the astronauts did not need to sit down in the ⅙-g environment near the Moon; they could stand in the LM cabin during descent and liftoff from the Moon. This became even more practical when the window layout was altered—the new small, angled windows worked better when the astronauts could stand closer to them, looking down. So the seats were pulled from the LM, saving more weight.

The overall structure of the LM was then considered. The top stage needed to be pressurized, which would require a certain amount of intrinsic strength. But the bottom stage of the lander—the octagonal structure that carried the descent engine, fuel tanks, and landing legs—was a different matter. Showing once again that they were shedding much of their traditional thinking, the engineers removed the metal panels from the sides of the LM descent stage, replacing them with a blanket of multiple layers of Kapton® (a type of Mylar®) and other films and fabrics that would provide even better thermal protection than aluminum sheets. It would even serve as a good micrometeorite shield. This shaved off many more pounds.

Hundreds of other parts of the design were scrutinized. The pressurized cabin of the LM, if built the same way space capsules were, would be far too heavy. So the Grumman engineers thinned the metal for the hull to dangerous levels, then added back just enough to do the job. The result was a pressure hull that was in many places only about as thick as a soda can, and in other areas little thicker than three layers of aluminum foil. On Earth, a dropped screwdriver would likely puncture it. The astronauts swore that when it was pressurized, the hull puffed out, and they took to calling the LM the "aluminum balloon." They were not far off.

Lightness was key to success. Consider the tyranny of the variables: Any weight added to the LM required that its fuel tanks had to be made larger, which made the craft heavier, which required larger fuel tanks, and so forth. Add to this that every pound of mass added to the LM resulted in adding 3 more pounds of fuel to the Saturn V. It quickly becomes a toxic feedback loop—the project soon felt like sharpening a pencil at both ends until one had nothing left to write with.

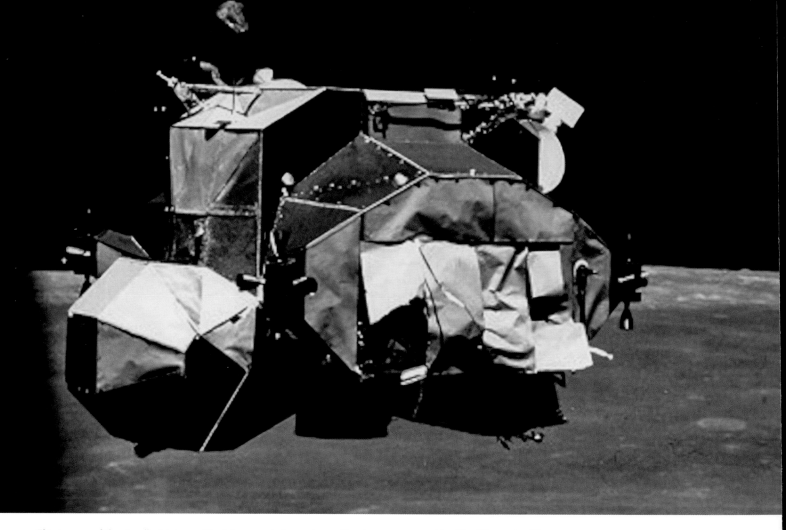

This image of the Apollo 16 Lunar Module ascent stage shows the damage caused by liftoff from the Moon.

Every part of the hull and fuselage was considered and thinned to the minimum amount that would hold the LM together. In the end, it all worked, but by a narrow margin.

In an eighteen-month period of the furious development cycle, some 50,000 drawings and schematics were created. As work proceeded, in came the NASA inspectors. They provided oversight at all the Apollo contractors, but with a company as small as Grumman, their presence was more keenly felt. An example of their detail-oriented supervision could go as follows:

Inspector: "Place support clamp PN AN 269972 on water line PN LDW 390-22173-3 on location shown in sketch . . ."

[Inspector shows technician a sketch.]

Technician: "OK, part number AN 269972 installed on water line LDW 390-22173-3."

Inspector: "Verify that rubber grommet on clamp is properly seated with no metal touching the tubing."

[Both look to see that this is the case.]

Technician: "Okay, rubber grommet on clamp is properly seated with no metal touching."

Inspector: "Align holes in clamp with hole in structure PN LDW 270-13994-I . . ."[25]

And it would all be checked again and again. That's how the work was done, and a big part of how success was assured.

By the end of 1965, the major components of the LM were coming together, but weight issues remained. The engineers had picked the low-hanging fruit, and now the weight reduction program went into overdrive. Kelly knew that excess pounds could kill the LM project, and he placed a literal bounty on excess mass. Two programs were initiated—one called Scrape and another called SWIP, the Super Weight Improvement Program. He would pay the Grumman workers a bounty for every ounce of weight they saved. This was available to any Grumman employee—if a janitor or secretary came up with a way to save enough weight, a cash reward went to them. And at $10,000 per pound (in 1960s dollars), it added up fast. All told, the weight reduction programs pruned about 2,500 pounds (1,135 kg) from the lander—just enough, and just quickly enough, to make the deadline.

DETAILS, DETAILS

A renewed scrutiny was then applied to every element of the LM, and this would continue through 1969. The

hull panels, landing gear, hatches, antennae—virtually anything structural was reexamined. Even the smallest parts were milled down in the shop until they broke under the stress of merely holding things together, then the machinist would remake the part, milling off just a wee bit less, and repeat the process, until the part was just strong enough. The engineers even resorted to "chemical milling," the removal of micrograms by dipping parts into chemical baths to eat away just a bit more of a given component.

Even the tiny pipes that supplied fuel to the maneuvering thrusters came under scrutiny. The pipe thickness was reduced, and the places where they were joined made lighter. But the engineers found that after these tubes sat for a time, filled with the corrosive

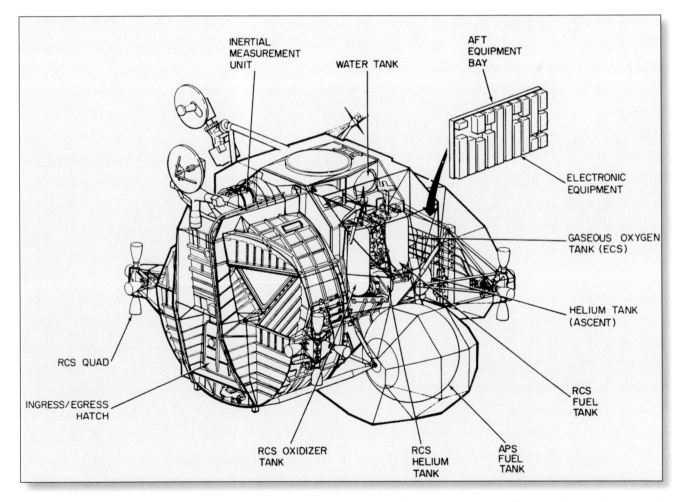

This schematic of the Lunar Module ascent stage shows the utilitarian nature of the craft—it was a hull wrapped around just enough space for two astronauts and the needed equipment. The ribbing on the pressure hull consisted of strips of metal affixed to the thin sheets of aluminum to give it added rigidity.

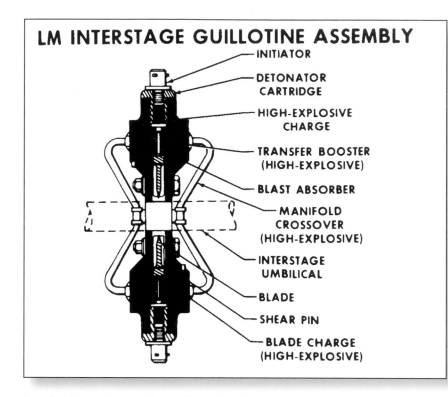

LM INTERSTAGE GUILLOTINE ASSEMBLY

- INITIATOR
- DETONATOR CARTRIDGE
- HIGH-EXPLOSIVE CHARGE
- TRANSFER BOOSTER (HIGH-EXPLOSIVE)
- BLAST ABSORBER
- MANIFOLD CROSSOVER (HIGH-EXPLOSIVE)
- INTERSTAGE UMBILICAL
- BLADE
- SHEAR PIN
- BLADE CHARGE (HIGH-EXPLOSIVE)

Detail of the explosive guillotine that, when fired, would chop in half the wiring bundle connecting the Lunar Module's lower and upper stages.

fuels used for the tiny maneuvering thrusters, the pipe joints, and sometimes the pipes themselves, began to leak dangerous fuel, allowing it to drip onto—and damage—other critical components. Right up until a few days before launch, Grumman engineers could be found tracing and fixing pinhole leaks in the system with tiny metal rings they would ever-so-carefully braze onto the delicate tubing. It was discovered that even a tiny residue of the detergent used to clean the components after fabrication was exacerbating this problem. It seemed to be a never-ending cycle.

As the weight reduction program continued, the LM was having total reliability engineered in. Everything had to be foolproof. When the astronauts were heading down to the lunar surface, they had to be able to throttle the engine in the descent stage. The readouts on the instrument panel would have to reflect the condition of the descent stage throughout the landing and afterward. But how do you engineer reliability into the connectors between two spacecraft stages? The obvious solution, and one used in spacecraft designs a few years earlier, would have been to design

reliable, detachable plugs where the ascent stage met the descent stage. That way when the time came to hit the ascent stage ignition button, the connectors would simply separate as the upper stage blasted toward orbit. This, however, also meant that the plug could short out or separate during flight if the LM experienced torqueing or extreme vibration—and the ride to orbit atop a Saturn V was not a gentle one. By their very nature, plugs and connectors can fail. There had to be a better way.

The solution was simple, though engineering it was not. The wire bundle between the two stages was about 4 inches (10 cm) in diameter where it passed through the ascent stage into the descent stage. The technicians ran continuous wires between the components in each stage—a wire would be soldered to a temperature sensor in the descent stage, for example, and run directly, without interruption, all the way from the connection in the landing stage to the gauge in the ascent stage instrument panel. There were hundreds of such wires.

In between the two stages, at the point where the wire bundle crossed from one stage to the other, the bundle was surrounded on two sides by sharp, thick blades with explosive charges packed behind them. Other explosives would be used to cut off the electrical current to those wires before separation; finally, four more charges would chop in half the bolts that held the two stages together. So when it came time to depart the Moon, all these explosive devices would detonate, in a carefully timed and staggered sequence of milliseconds, to send the astronauts toward their orbital rendezvous with the waiting Command Module and their trip home.

The wiring itself was another area of concern—copper wire is heavy, and there was a lot of it in the LM. The engineers reexamined their assumptions about what was necessary to get the job done, and a redesign of the grounding wires alone, critical to completing electrical circuits, saved almost 60 pounds (27 kg).

This is just one of many examples of systems engineered into the LM to assure that reliability was as close to 100 percent as was possible.

Despite the fact that weight was being shed and problems solved, the LM program fell further and further behind schedule—one NASA quality control review alone listed over 300 areas of critical concern in the fabrication of the LM. By 1968, with Apollo 8 ready to fly and the first lunar landing mission just a few months after that, there was still no LM ready that was light enough to fly. With everything else ready to go, the pressure was on at Grumman—its lunar lander program might be the only thing that forced the first landing beyond Kennedy's deadline.

The long hours continued, including vast amounts of overtime and weekend shifts, causing stress to the point that some of the employees began to crack.

A Lunar Module being assembled by Grumman workers. This particular LM was a test unit for use on the ground, but it was basically identical to the flight units.

Kelly instituted what he called "Charm School" in an attempt to evaluate their physical condition and aptitude to endure prolonged periods of pressure, and to help them learn to cope or switch them to other less stressful assignments. In addition, Kelly spent more and more time walking the shop floor, trying to spot and solve problems as well as keep an eye on the well-being of his workforce.

The months rolled on with little letup in the grinding schedule. Problems continued to appear and were mercilessly pounded into submission until, early in 1969, Grumman at last had a flight-worthy Lunar Module. The costs—both in dollars and in labor—had been staggering, but it all came together at the eleventh hour. In the end, nine Lunar Modules flew with crews,

eight of them out to the Moon. One lander, LM #7, was particularly remarkable. It did not land on the Moon but instead saved the crew of the stricken Apollo 13 mission. On the way to the Moon, the Command Module was crippled by an explosion early in the flight, and the crew's salvation came in the form of their Lunar Module, which provided them with the rocket engine and life support that allowed them to return home alive by the narrowest of margins. The LM proved its mettle under fire, and it did so brilliantly.

In the end, the Lunar Module was the only major component of the Apollo system that never suffered a noteworthy malfunction in flight—a testament that Tom Kelly, and his thousands of associates at Grumman, could be proud of.

T-MINUS ZERO

"OKAY, HOUSTON. YOU SUPPOSE YOU COULD TURN THE EARTH A LITTLE BIT SO WE CAN GET A LITTLE BIT MORE THAN JUST WATER?"

—Mike Collins, Apollo 11 Command Module pilot

THE APOLLO PROGRAM WAS IN HIGH gear now—Apollo 8 had orbited the Moon ten times in December 1968, Apollo 9 had extensively tested the Lunar Module in Earth orbit in March 1969, and Apollo 10 had barnstormed the Moon in May. For that mission, Tom Stafford and Gene Cernan (both of whom would fly on later Apollo missions) had climbed aboard the LM and descended toward the lunar surface in a non-landing test. At about 50,000 feet (15,250 m) altitude, they staged the LM, separating the ascent stage from the descent stage, and then rode the ascent stage back to orbit and rendezvous with the Command Module. Other than some gyrations after staging, which were quickly dealt with, the flight had been by the book. All the major components of the Apollo spacecraft had been tested at least once—it was now time for the first landing attempt, and that fell to the crew of Apollo 11.

At a July 5 press conference, the astronauts assigned to the Apollo 11 flight responded laconically to last-minute questions from reporters. Of the thirty-seven questions, Armstrong answered all but ten

OPPOSITE: The Apollo 11 Saturn V as seen just before dawn.

ABOVE: The Lunar Module ascent stage returns to orbit to dock with the Command Module during the Apollo 10 mission.

A preflight press conference for Apollo 11. From left, Aldrin, Armstrong, and Collins.

of them. Some of the questions were rather trivial, for example, asking how the Command Module came to be named *Columbia*. "Columbia is a national symbol," Armstrong responded, "and, as you all know, it was the name of Jules Verne's spacecraft that went to the Moon."

"Was Verne one of your favorite authors?" a reporter asked. "No, I don't think so, but I had certainly read the book," Armstrong replied in dry tones. More to the point, another reporter asked him, "Have [you] decided on something suitably historical and memorable to say when you perform this symbolic act of stepping down on the Moon for the first time?"[26]

Uh-oh. This had been an item of some discussion, as it would reflect the energies of a nation for the better part of a decade. There was a lot at stake, and insofar as anyone knew, a NASA public relations team had been busy creating the perfect words to be spoken at that historic

moment. But this was not the case—this decision had been left up to Armstrong. As the issue of what to say on the Moon got batted around NASA prior to the mission, the head of public relations, Julian Scheer, had written a memo that stated, in effect, that since Queen Isabella had not specifically instructed Christopher Columbus on what he would say when he encountered land during his journey to the New World, NASA would not be telling Armstrong what to say when he stepped onto the Moon. It had become a sensitive point, though you would not guess it from Armstrong's response to the reporter: "No, I haven't." And that was that.

Other reporters asked about what symbolic items would be left on the Moon, beyond the US flag, and if there was any potential legal significance to this. Armstrong replied, "I think we might refer you to the plaque again," referring to the plaque affixed to the front leg of the Lunar Module. "It says, 'We came in peace for all mankind.' I think that is precisely what we mean." Issue closed.

Finally, a reporter asked Armstrong, "What would, according to you, be the most dangerous phase of the flight of Apollo 11?" Armstrong was not baited into a dramatic response. "Well, as in any flight, the things that give one most concern are those which have not been done previously, things that are new," he began. "The LM engine must operate to accelerate us from the Moon's surface into lunar orbit, and the Service Module engine, of course, must operate again to return us to Earth." He added, "As we go farther and farther into spaceflight, there will be more and more of the single-point systems that must operate." He closed the answer with, "We have a very high confidence level in those systems, incidentally." Ask an engineer an emotionally based question, and that's often the kind of answer you will get.

There had been concerns about the LM's ascent engine firing system, incidentally, with Armstrong quietly inquiring at one point about backup methods for starting the engine. It was a very simple machine, but it had only one starting system. There were two pressurized fuel tanks (meaning that there were no turbopumps to fail) with hypergolic fuels (they explode on contact, no igniter system needed), and that were operated by two small valves that opened with explosive charges. Press the button and—*bang!*—off you went to lunar orbit. But what if the electrical jolt that caused the explosives to fire did not operate properly? Couldn't the engineers add two hand-operated valves to the engine to ensure that the astronauts could leave the lunar surface? The answer was no—it would add too much weight—and the engineers at Bell Aerosystems, the contractor for the LM ascent engine, assured them that the system was fail-safe. In the end, Armstrong agreed, though possibly with some reluctance.

TIME TO GO

On the morning of July 16, chief astronaut and compatriot Deke Slayton awakened Armstrong, Aldrin, and Collins at 4:00 am. They were sleeping in crew quarters at the Kennedy Space Center and had been effectively quarantined for days before the flight—nobody wanted a case of the sniffles, or worse, to strike a crewmember on the way to the first Moon landing. Last-minute medical checks were made, and then the three men were off to breakfast. They ate the traditional preflight astronaut fare: steak, eggs, toast, fruit juice, and coffee. The menu was partly a matter of tradition but also represented a low-residue diet that would minimize bowel movements during the flight—doing a "number two" inside the Apollo spacecraft in weightless conditions was an unpleasant affair involving a special plastic bag with a circular adhesive seal and a built-in glove to mix the feces with germicide. Nobody liked it.

Slayton and fellow Apollo astronaut Bill Anders joined the crew at breakfast. Then, about a half hour later, the astronauts went upstairs to suit up and prebreathe pure oxygen in preparation for launch, a few hours hence.

Soon the trio made their way to the van that would drive them the 8 miles (13 km) from the crew quarters

The prelaunch breakfast for Apollo 11. Bill Anders is seen at the far left, then Armstrong, Collins (in foreground), Aldrin, and Slayton (holding chart).

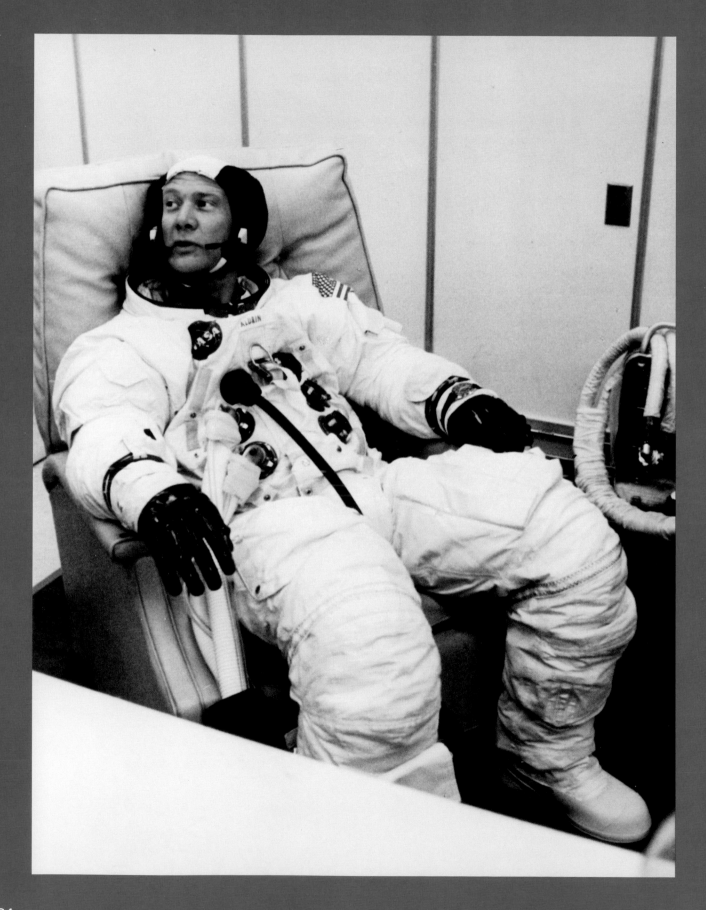

OPPOSITE: Buzz Aldrin sits in a lounge chair after being suited up, awaiting his helmet to pre-breathe pure oxygen before launch.

ABOVE: Mike Collins during suit donning, doubtless keeping the mood light.

to the launch pad. The rocket stood before them, awash in the dawn light, with vapor trails streaming from the vent ports high above. Looking at the sight, Collins reflected, "The two partners make quite a contrast, the rocket sleek and poised and full of promise, the tower old, gnarled, ungainly, and going nowhere."[27]

As the astronauts rode the elevator to the Command Module, some 340 feet (103 m) above the ground, astronaut Fred Haise was completing a 417-step checklist inside the capsule. Every switch had to be in the proper position, and every gauge and indicator light had to show the proper reading.

The three crewmates were helped into *Columbia* by the pad team and seated in their assigned couches. Dressed in their bulky suits, their shoulders were touching. Once en route to the Moon, they would slip out of the pressure suits, and in the zero-g environment

the capsule would seem much larger. For now, it was cramped. "Nearly every available cubic inch of space has been used, save for two great holes in the lower equipment bay which are reserved for the boxes of Moon rocks to be brought back from the LM," Collins later recalled.[28]

As they settled into the capsule and began their final checks, over a million people jostled for the perfect—or last available—viewing spot all around the Cape. There was a minimum distance of about 3 miles (5 km) between the rocket and the closest viewing area for non-launch related personnel—that is where many members of the press and the VIPs were located. That was also considered the minimum distance where NASA could assure the safety of unsheltered viewers if the rocket blew up during launch, as it would have had the explosive force of a small nuclear weapon. While the astronauts were not dwelling on such thoughts, the emptiness of the launch complex was apparent to Collins. Before entering the rocket, he recalled thinking: "Something seems wrong, and I suddenly realize what it is. The place is deserted! . . . It seems as if some dread epidemic has killed all but those protected by pressure suits."[29] Once inside the capsule, as they got busy preparing for launch, this was forgotten.

Down in the Firing Room, where the rocket was being carefully monitored, technicians were joined by the senior members of the Apollo team. Rocco Petrone, the director of launch operations, sat on a raised dais, surrounded by launch controllers looking intently at their consoles. In a glass enclosure nearby were Wernher von Braun; George Mueller, who now ran the Office of Manned Space Flight; and Air Force General Sam Phillips, the director of the Apollo program. They chatted amongst themselves as the clock ticked inexorably toward T-minus zero.

As the count neared the final minutes, the crowds both inside the Firing Room and outside around the Cape grew increasingly quiet. The large, lit numerals on the countdown clock in the press area ticked down in reverse.

In the Command Module, Armstrong quietly moved his hand to the abort handle, a T-shaped unit that sat on the arm of his couch. If anything went wrong during launch, all he had to do was twist the

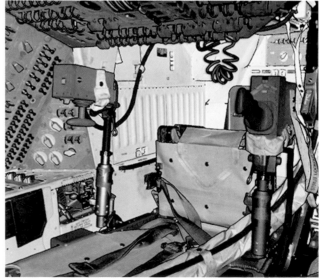

TOP: The Kennedy Space Center Firing Room, from which the Saturn V rocket was monitored until it cleared the launch tower.

BOTTOM: On July 16, 1969, all available space around the Cape was filled with spectators awaiting the launch of Apollo 11. In the VIP section were former president Lyndon Johnson and then-current vice president Spiro Agnew, seen here watching the ascent of Apollo 11.

ABOVE: Interior of the commander's seat in the Command Module. To center left, at the end of the armrest, is the T-shaped abort handle. In the event of a launch emergency, a quick twist of this handle would have initiated the separation of the Command Module from the Saturn V. Fortunately, it was never needed.

handle and the emergency escape system, a set of rockets that sat atop the Command Module, would whisk the astronauts away from a malfunctioning booster. It was a last resort, and not something they liked to think about (an abort was never tried with a crew as it was considered quite dangerous), but they all knew it was there. As if to quell any hesitation, current NASA administrator Tom Paine had said just two nights prior, "If you have to abort, I'll see that you fly the next Moon landing flight. Just don't get killed."[30] It was his way of saying, "Don't wait too long to abort if things go haywire."

"LIFTOFF! WE HAVE A LIFTOFF!"

As the count neared zero, Aldrin, in the center seat, recalls turning first to Armstrong, then to Collins, with a wide grin. After years of preparation and effort, they were actually going to the Moon.

The voice of Jack King, NASA's public relations man at launch control, was heard across the Cape.

"Ten, nine, ignition sequence start . . ." King said.

The main engines powered up, with flames leaping toward the deflection trough below the launch pad and billowing out to the side. The fuel-heavy Saturn would devour 23 tons (21 metric tons) of kerosene and liquid oxygen before it left the ground.

King continued: "Six, five, four, three, two, one, zero . . . all engine running." (He clipped the s off of "engines" in the excitement.) "Liftoff! We have a liftoff, thirty-two minutes past the hour, liftoff on Apollo 11."

In the background of launch control, you could hear a controller's voice say something, which King repeated—"Tower cleared!" The Saturn V had lifted past the launch tower. This was important, as some people had lingering concerns that if the rocket bobbed or swayed too far, it could hit the launch tower, with potentially disastrous results. As Collins later commented, "For the first ten seconds we are perilously close to that umbilical tower. I breathe easier as the ten-second mark passes."[31]

Armstrong made his first transmission from the ascending Saturn V: "We got a roll program." The rocket was rolling onto its side to head off at an angle toward the equator.

TOP: The launch of Apollo 11 on July 16, 1969.

BOTTOM: Jack King, the "voice of launch control" during the Apollo missions.

Paul Haney, the chief of public affairs in Houston, took over the commentary. "Neil Armstrong reporting their roll and pitch program, which puts Apollo 11 on a proper heading."

The Saturn V continued to ascend, the crew enduring the violent shaking of the first stage. Despite the best efforts of the engineers at Rocketdyne to tame the F-1 engine, the two-plus minutes that the first stage burned were still a jolting ride.

While Haney kept the public informed of the mission's progress, fellow astronaut Bruce McCandless was the CAPCOM for the launch. His calm voice kept up a running dialog with the astronauts. After about two and a half minutes, the five F-1 engines shut down, and the rocket prepared to stage.

"Apollo 11, this is Houston," McCandless said. "You are go for staging."

"Inboard cut-off," responded Armstrong. The final F-1 engine had stopped firing, and they were now coasting. Then, almost blandly, "Staging."

The first stage separated explosively from the second stage and fell to the Atlantic Ocean, far below. Soon thereafter, the S-II stage's five J-2 engines lit up.

"And ignition," Armstrong added moments later. McCandless replied, "Thrust is go, all engines. You're looking good."

The rocket continued on into orbit, with three grinning astronauts on their way into history. On the ground in Florida there were backslaps and smiles all around, especially in the VIP area of Launch Control, where von Braun and other brass from NASA had assembled to watch the launch. This was the fifth successful launch of a Saturn V; they were nonetheless relieved as it neared orbit.

At just over nine minutes into the flight, the S-II's engines were shut down, and the rocket staged again. The single J-2 rocket on the S-IVB ignited, burning for

OPPOSITE: One of the most iconic shots of the space race, Apollo 11 ascends from Cape Canaveral, with the American flag in the foreground.

ABOVE: Wernher von Braun is seen on the left, with binoculars, then George Mueller and Sam Phillips at their positions in the Apollo Firing Room shortly after launch.

a shade over two more minutes and placing Apollo 11 into the proper orbital trajectory. At eleven minutes and forty-two seconds into the flight, Armstrong confirmed that the first burn of the S-IVB was complete.

"Shutdown," Armstrong relayed.

Some technical transmissions, then:

"Apollo 11, this is Houston," said McCandless. "You are confirmed *go* for orbit."

At about two hours and forty-four minutes, after making sure that everything was shipshape and that the spacecraft was ready to continue its journey, the crew initiated the trans-lunar injection (TLI) burn with the restartable J-2 engine. Six minutes later, the single engine shut off—it was the last time it would fire.

FETCHING THE LUNAR MODULE

Just under a half hour later came one of the trickier parts of the mission besides the lunar landing itself. Called the Transposition and Docking Maneuver (TDM), this involved separating the Command Module (CM) from the S-IVB stage, pulling ahead of it,

The Lunar Module sits exposed, awaiting docking and removal from the S-IVB stage by the Command Module. This shot is from Apollo 9.

turning 180 degrees, then thrusting back toward the Lunar Module, seated neatly atop the S-IVB stage with its legs tucked in along its sides. The four panels that had enclosed it during the wild ascent into orbit had unclamped and peeled back like flower petals, drifting away into space, leaving the LM exposed, the gold-foil covering on the descent stage glinting in the sunlight.

Using ranging radar and an optical sight—this was 1969 after all—Mike Collins would utilize the skills he had sharpened during those hundreds of hours in the simulator to slowly, ever so slowly, fly to the LM and dock with it. This was just one more critical step in the path to the Moon, but a big one. If he came in

at the wrong angle, or at just slightly too high a rate of speed, he could puncture or crumple the incredibly delicate LM.

On the front of the Apollo capsule was a hatch with a docking probe on it—it looked like a cone made of metal sticks. Atop the LM was a hatch of the same size with a tapering funnel inside. The trick was to slide the docking probe inside the funnel with just enough force for it to align itself and snug the two spacecraft together, then docking latches would snap closed, securing the two spacecraft in an airtight embrace. At three hours and fifteen minutes into the flight, Collins separated the CM from the S-IVB. He pulled ahead of the upper stage, about 100 feet (30 m) by his estimate, then rotated the CM 180 degrees, and proceeded to thrust slowly back toward the LM hatch directly ahead. This was a largely manual maneuver, with Collins sighting through an optical rendezvous sight. As written by the commander of a later mission, Dave Scott, "[This] is pretty much a manual out-the-window operation, and the actual thrusting time, maneuver, and etcetera are left to the pilot. The checklist here is more of a guideline rather than a rigid set of procedures as in other situations. Much like the lunar landing, the pilot in the left seat flies mostly visually using the outside scene for reference."[32]

Collins closed the gap between the two spacecraft carefully, with his crewmates watching the acceleration indicator on the instrument panel. Collins later noted that through his eyepiece, the LM looked like "a mechanical tarantula crouched in its hole. It's one black eye peers malevolently at me."[33] The astronauts kept up a running commentary as they closed on the S-IVB.

Armstrong: "Beautiful."

Aldrin replied, "It really looks nice, doesn't it?"

Collins added: "Hey, we're closing in a leisurely fashion."[34]

After some more technical transmissions, Collins said: "Stand by; we're closing."

The docking probe of the CM slid into the receiver on the roof of the LM with a squeaky groan. There was no way for the direction of the two spacecraft to be perfectly aligned, so Collins waited a bit for the two spacecraft to settle before activating the docking latches. A few minutes later, the latches clamped shut and the CM was firmly docked to the LM. Collins began to immediately worry about how much of their precious maneuvering fuel he had used.

"That wasn't the smoothest docking I've ever done," he noted with a bit of frustration. Armstrong replied, "Well it felt good from here." But Collins was unhappy with the fuel consumption: "I mean the gas consumption would be a lot more than I would have guessed, you know? I thought I could about equal the simulator in . . . and I didn't—I bet you I used—I hate to quote a number, but I've been down around 30-some pounds [14 kg] in the simulator, and I'll bet this was 50, 60 pounds [23, 27 kg], something like that."

They reported their status to Mission Control and within ten minutes were pressurizing the LM. A bit less than an hour later, Collins was again in the driver's seat, where he armed his systems and pulled the LM free of the S-IVB.

GOODBYE, S-IVB

A short time later, the S-IVB stage ignited a small auxiliary rocket motor to push it off axis from the CM to prevent an inadvertent collision—now that they had the LM in tow, the last thing the crew aboard Apollo 11 wanted was to reencounter the S-IVB stage. It had done its job. The "Evasive Maneuver" was executed,

The Apollo docking system in test. At bottom is the pyramidal arrangement that made up the Command Module's docking probe, and at top the funnel-shaped receiver that sat atop the Lunar Module. Once the two were engaged, a series of clamps closed to secure the spacecraft together. Then the mechanism would be manually removed by Collins from inside the Command Module to allow Armstrong and Aldrin access to the Lunar Module.

THIS IMAGE IS ASSEMBLED from multiple photos taken during the Moonwalk, far from the Lunar Module. The handle to the left is from the Apollo Lunar Surface Close-up Camera (ALSCC), which would be left behind when Armstrong and Aldrin departed. This somewhat cumbersome device had a long handle affixed to it to allow the astronauts to get close-up pictures of the lunar surface without bending over,

which was impossible in their stiff, bulky Moonsuits. The shallow crater to the right is about 40 feet (12 m) in width, and the curvature of the surface is somewhat exaggerated by the photo. This was about as exciting as the Apollo 11 landing site got in terms of terrain—it was deliberately chosen to be safe (a geologist might say boring), while still providing enough objects of interest to make the visit worthwhile.

This image from Apollo 9 shows the same view Collins had from the Command Module as he slowly closed on the Lunar Module for docking. The funnel-shaped docking receiver can be seen at the center top of the LM. Credit: NASA.

and the S-IVB stage was sent off into a large orbit around the Sun.

A bit later, Armstrong relayed what he was seeing out the window to Mission Control.

"We didn't have much time, Houston, to talk to you about our views out the window when we were preparing for LM ejection; but up to that time, we had the entire northern part of the lighted hemisphere visible including North America, the North Atlantic, and Europe and Northern Africa. We could see that the weather was good all—just about everywhere. There was one cyclonic depression in northern Canada, in the Athabasca—probably east of Athabasca area. Greenland was clear, and it appeared to be we were seeing just the icecap in Greenland. All North Atlantic was

A period illustration from NASA press materials showing the Command Module, now docked with the Lunar Module, firing thrusters to back away from the S-IVB stage.

RIGHT: The crew of Apollo 11 checks out their Command Module prior their mission. While they talked about "moving around the capsule" during flight, the small interior space of the spacecraft is apparent.

BELOW: In this modern graphic, the Command Module and Lunar Module are seen docked, as they looked in transit to the Moon. The duo rolled along their long axis to equalize the heat from the sunlight striking the exposed hull.

pretty good, and Europe and Northern Africa seemed to be clear. Most of the United States was clear. There was a low—looked like a front stretching from the center of the country up across north of the Great Lakes and into Newfoundland."

Then, with typical dry wit, Collins added, "*I didn't know what I was looking at, but I sure did like it.*"[35]

About five hours later, Collins initiated another maneuver, this one to put the CM-LM combination into a slow roll around its long axis, at the rate of about one revolution per minute. This was called Passive Thermal Control, or PTC, but was informally called "barbecue mode." The purpose of rolling the spacecraft was to equalize the temperature across the hull. It was a kind of ad hoc cooling system that negated the need for large, bulky devices to cool the spacecraft—without this roll, one side could rapidly heat to over 250 degrees Fahrenheit (121°C), while the other would plunge to a couple hundred below zero. It was basic, but it worked.

This completed, they could finally get out of their pressure suits (they had been wearing them since launch), clean up, stow loose equipment, and relax for a bit. Armstrong, Aldrin, and Collins were on their way to the Moon at 10,208 miles per hour (16,428 kph).

CHAPTER 9

ARRIVAL

"COME ON NOW, BUZZ, DON'T REFER TO [CRATERS] AS
'BIG MOTHERS'; GIVE THEM SOME SCIENTIFIC NAME."

—Mike Collins, Apollo 11 Command Module pilot

THE MISSION PROCEEDED SMOOTHLY as they coasted to the Moon. There was time to eat, rest, and even send a first TV transmission back to the millions of people following the mission on Earth. At about twenty-six hours into the flight Collins once again settled into the pilot's seat to perform the midcourse correction burn—this was a firing of the CM's rocket engine to adjust their trajectory to line them up for lunar orbit. Prior to this, Collins had to perform a waste-water dump, sending excess water overboard through a small valve. This was done before the corrective burn to compensate for any propulsive effects of the spray of water leaving the Command Module—even that tiny jet of vapor could affect the trajectory. With the water dump complete, Collins used a small periscope to align the spacecraft properly in relation to the stars. Even with all their advanced (for the time) technology, celestial navigation, an art honed by Earth-bound mariners over centuries, was part of the drill. In this case, however, rather than finding their position on the two-dimensional surface

of the ocean, they were determining their position in three-dimensional space. Once this was figured out, Collins fired the CM's engine, and they were pointed exactly where they wanted to go. With just a few seconds of thrust, they had adjusted their ultimate orbit around the Moon from about 201 miles (323 km) above the surface to 69 miles (111 km).

At sixty-one hours into the mission, Apollo 11 passed an invisible point in space called the *equigravisphere*, where the gravitational pulls of the Earth and Moon are equal. From here on out, the crew would accelerate toward the Moon as its pull yanked them in. Precision was critical.

Fourteen hours later, they were closing in on the Moon. The crew strapped in to their couches in preparation for the engine burn that would "put on the brakes" and insert them into that perfect lunar orbit they were aiming for. For this, the CM-LM was oriented with the rocket engine pointed in the

OPPOSITE: The lunar far side as an Apollo capsule passed behind the Moon. This view is only possible from lunar orbit.

same direction they were traveling. The spacecraft slowly swung around what Aldrin referred to as the "left-hand side of the Moon," where they would pass behind it and out of view of the Earth. That also meant that they would be out of radio contact, since the rocky bulk of the Moon blocked their signal. It was then that they would fire the rocket engine. Mission Control would not know if the engine firing had been successful until Apollo 11 emerged from the other side of the lunar disk.

As they neared the lunar limb (the edge of the Moon), a last radio exchange with Mission Control: "One minute to LOS. Mark that," said Aldrin. (LOS being short for *Loss of Signal*, indicating the moment Apollo 11 would pass behind the Moon and the radio transmissions would drop out.)

Mission Control replied, "Apollo 11, this is Houston. All your systems are looking good going around the corner, and we'll see you on the other side. Over."

As they passed across the backside of the Moon, Aldrin marveled at the roughness of its surface. The Moon is "tidally locked" with the Earth, meaning that one side forever faces our planet. The backside (sometimes inaccurately referred to as the *dark side*; it in fact goes through the same phases as the near side) is forever hidden from human eyes—unless seen from the point of view of a spacecraft.

"The back side of the Moon was much more rugged than the face we saw from Earth," Aldrin recalled, "This side had been bombarded by meteors since the beginning of the solar system millions of centuries ago."[36]

Collins initiated the firing of the CM's engine by entering codes into the onboard computer interface— the DSKY (short for "display/keyboard"). The thrust slowed them enough to fall into lunar orbit—otherwise, they would have shot past it. In the latter scenario, if the CM's engine had failed to ignite when they were behind the Moon, they would have continued into space. Prior to the correction burn hours before, had the CM's engine failed, Apollo 11 would have entered what is called a "free return" trajectory, returning to Earth without further intervention. But once the midcourse trajectory burn was completed, that option was no longer there. Fortunately, fire the engine did, and the burn lasted about six minutes, braking them into lunar orbit.

When the engine shut down, Aldrin checked the instruments to ascertain how close the orbit was to their expectations. Collins chuckled as he looked on, commenting on their estimated 60-mile-high (97-km) orbit: "Well, I don't know if we're 60 miles or not, but at least we haven't hit that mother."

Then Aldrin read off the altitude indicator: "Look at that! Look at that! 169.6 by 60.9!"

Collins replied, "Beautiful, beautiful, beautiful, beautiful!" Then, "You want to write that down or something? . . . Write it down, just for the hell of it. 170 by 60, like gangbusters!"

Then Collins turned to look out the window. "Hello, Moon. How's the old backside?" Thirty minutes later, they cleared the Moon's limb, and Mission Control crackled in the headphones.

"Apollo 11, Apollo 11, this is Houston. Do you read? Over."

Aldrin responded, "Yes, we sure do, Houston. The LOI-1 [Lunar Orbital Insertion-1] burn just nominal as all getout, and everything's looking good."

Collins was beaming. He had delivered his comrades to lunar orbit—now it was up to them to cross the last 60 miles (97 km) to the rocky, pitted surface below.

THE LAST LAP

About four days into the mission, approximately ninety-five hours after launch, Armstrong and Aldrin entered the Lunar Module for final preparations for the lunar landing. They were in their eleventh orbit, and if all went according to plan would land within hours.

On the ground in Houston, a new shift was arriving at the Manned Spaceflight Center. This was Flight Director Gene Kranz's "White Team," the group that would guide the two astronauts to the Moon's surface. Among them were:

- Gene Kranz, flight director during the first lunar landing.
- Steve Bales, guidance officer (GUIDO), expert on the Eagle's guidance system.
- Sy Liebergot, electrical, environmental, and communications officer (EECOM).

- Chuck Dietrich, retrofire officer (RETRO), who would oversee abort options during the landing.
- Jack Garman, group leader, program support group, Apollo guidance software and computer expert.
- Charlie Duke, one of the most knowledgeable astronauts when it came to the LM, whom Armstrong wanted as the CAPCOM during the landing.
- Bob Carlton, LM descent engine specialist (CONTROL).
- Joe Gavin, director of the Lunar Module Program for Grumman, along with Tom Kelly and others from Grumman, who were there to oversee the performance of their landing machine.

Kranz remembers driving to work for the shift that would land the first Americans on the Moon. "I had my fresh haircut, and my wife had packed me a sack lunch that was enough for three shifts of people. As I arrived at my parking spot at [the Manned Space Flight Center], I realized I didn't remember driving through Clear Lake, or anything else."[37] Kranz was intently focused on what he was about to oversee.

Joe Gavin recalled, "The whole thing was tense, because we were basically aircraft designers. In the aircraft business you always flight tested something before you delivered it. In the case of the Lunar Module, you couldn't flight test it. Every launch was a brand-new vehicle."[38]

Steve Bales was tense for other reasons: "When we came in that morning, the Lunar Module was dead." He meant that the systems were powered down and needed to be turned back on prior to landing. "We had to power it up, get the thing aligned and checked out. In the simulations, that's where we'd always had the biggest difficulty, really. We had never completed [this in training] without some major problem."[39]

The team members arrived, got their coffee, and the smokers among them positioned their ashtrays on their consoles. Once prepared, they took over from the outgoing shift. It was going to be a long but exciting day—regardless of the outcome.

Just two days before this, in Washington, President Richard Nixon's speechwriter had put the finishing touches on a "just in case" speech to be used if something went wrong on the lunar surface. It had been prepared at the suggestion of Apollo 8 astronaut Frank Borman, who was working as a NASA liaison with the White House. It read:

Fate has ordained that the men who went to the Moon to explore in peace will stay on the Moon to rest in peace.

These brave men, Neil Armstrong and Edwin Aldrin, know that there is no hope for their recovery. But they also know that there is hope for mankind in their sacrifice. These two men are laying down their lives in mankind's most noble goal: the search for truth and understanding.

They will be mourned by their families and friends; they will be mourned by their nation; they will be mourned by the people of the world; they will be mourned by a Mother Earth that dared send two of her sons into the unknown.

In their exploration, they stirred the people of the world to feel as one; in their sacrifice, they bind more tightly the brotherhood of man.

In ancient days, men looked at stars and saw their heroes in the constellations. In modern times, we do much the same, but our heroes are epic men of flesh and blood. Others will follow, and surely find their way home. Man's search will not be denied. But these men were the first, and they will remain the foremost in our hearts.

For every human being who looks up at the Moon in the nights to come will know that there is some corner of another world that is forever mankind."[40]

Fortunately, this moving speech was never needed. Instead of being read to a grieving nation, it was immediately consigned to an administration filing cabinet once the landing had been successfully executed.

To : H. R. Haldeman

From: Bill Safire   July 18, 1969.

IN EVENT OF MOON DISASTER:

Fate has ordained that the men who went to the moon to explore in peace will stay on the moon to rest in peace.

These brave men, Neil Armstrong and Edwin Aldrin, know that there is no hope for their recovery. But they also know that there is hope for mankind in their sacrifice.

These two men are laying down their lives in mankind's most noble goal: the search for truth and understanding.

They will be mourned by their families and friends; they will be mourned by their nation; they will be mourned by the people of the world; they will be mourned by a Mother Earth that dared send two of her sons into the unknown.

In their exploration, they stirred the people of the world to feel as one; in their sacrifice, they bind more tightly the brotherhood of man.

In ancient days, men looked at stars and saw their heroes in the constellations. In modern times, we do much the same, but our heroes are epic men of flesh and blood.

READY IN MISSION CONTROL . . .

Kranz stood before his team, steely eyed and intense. His bearing was ramrod straight, a former Marine with a close-chopped crew cut topping off his medium frame. He was rugged, intense, and relentlessly professional. The sum total of these traits was a man perfectly bred for this moment. Kranz was just thirty-six years old.

He had been with NASA since the Mercury program, working under the legendary Chris Kraft, now the director of flight operations at the Manned Spacecraft Center. Kraft had been with NASA since the beginning, and he had assigned Kranz to write the flight procedures books. Kranz knew those rules cold.

Kranz had trained his people—he had practically lived with them for the past seven years—and he trusted them implicitly. The average age of his controllers was just twenty-six. "I wanted people young enough to not know failure," he said.[41]

He stood before his team, resplendent in his new vest—his wife, Marta, made him a bespoke vest before each flight. These ranged from plain white to gaudy red-white-and-blue stripes, but each was magnificent in

its own way. Today, the vest he wore was white brocade with silver accent thread—one of the subtler ones. He stood before his team and pulled the vest smooth over his Marine-trim physique. He had prepared a short speech for his flight control team. It went something like this:

"Okay, flight controllers, listen up. From the day of our birth, we were meant for this time and place, and today we will land an American on the Moon. Whatever happens here today, I will stand behind every decision you will make. We came into this room as a team and we will leave as a team. And from now on, no person will enter or leave this room until either we have landed, we have crashed, or we have aborted. Those are the only outcomes from this time on."[42]

To him, the stakes were high and the outcomes simple. Of his description of the results of that day, he later said, "The last two outcomes were not good."[43]

And with that, he ordered the doors to Mission Control locked and the circuit breakers in the power supply locked in place. "I was leaving nothing to chance that I had control over," he said.[44]

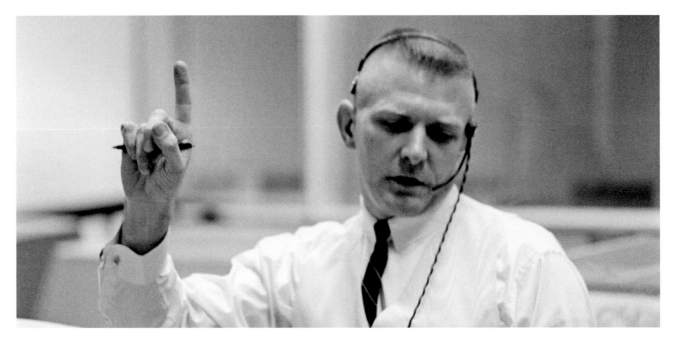

OPPOSITE: A "just in case" speech prepared by William Safire for Richard Nixon in the event that disaster befell the Moonwalkers of Apollo 11.

ABOVE: Gene Kranz with his signature crew cut, seen here during the flight of Apollo 5.

After undocking, Armstrong performed a slow roll in front of Collins to allow him to check the condition of the Lunar Module prior to initiating their lunar descent.

. . . AND IN SPACE

While Armstrong and Aldrin went through their prelanding checklists inside the Lunar Module *Eagle,* Collins fretted in the Command Module *Columbia.* Though he had not told his companions, he gave them a fifty-fifty chance of success—though what ultimate outcome he foresaw was open to interpretation. "You cats take it easy on the lunar surface," he said.[45] *Columbia* and *Eagle* were still docked, flying in formation above the Moon.

A few minutes later, at one hundred hours, twelve minutes into the mission, Collins hit the switch that released the *Eagle.* The docking mechanism had a spring that gently pushed them apart. That was a known value and had been factored into the landing trajectory—like the midcourse water dump a few days before, every little detail had to be included in the calculations. What was *not* factored in was the tiny residual air pressure in the tunnel between the CM and

LM. It was just enough to give the *Eagle* a bit of extra boost, which would have immense ramifications for their landing later.

Armstrong used the thrusters to maneuver away from *Columbia,* then slowly pirouetted in front of Collins to allow for a visual scan of the LM. Everything looked good to Collins—most importantly, there was no damage to the LM and all four landing legs were extended and locked.

"I think you've got a fine-looking flying machine there, *Eagle,* despite the fact you're upside down," he said.

Armstrong replied, "*Somebody's* upside down."

"You guys take care," Collins added.

Armstrong replied, simply, "See you later," as if he were heading down to the local watering hole for a drink rather than preparing for the first lunar landing. But that was his way.

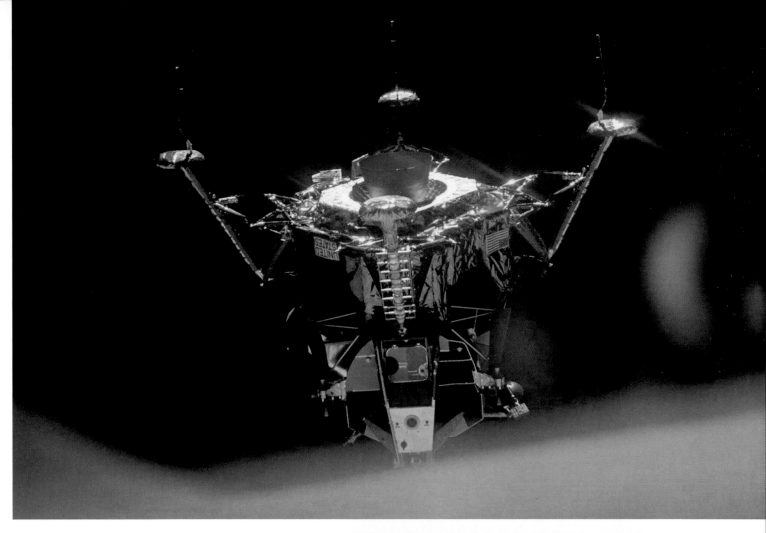

ABOVE: The Lunar Module *Eagle* just prior to lunar descent.

RIGHT: Armstrong as he would have appeared preparing to descend to the Moon—he would put on a helmet before the descent. Image taken in the Lunar Module simulator shortly before the flight.

Using the sight on his navigational telescope, Collins watched his companions recede into the distance as they headed downrange to initiate their first engine burn that would slow them enough to descend. He watched until they were about 100 miles (160 km) away, and the speck he knew as *Eagle* faded out. For the next twenty-four hours he would be flying solo, the most solitary man in human history.

For the next two hours and twenty minutes, *Eagle* continued to orbit as the two men inside went through their checklists and prepared to commit to the great task they had trained to do for eighteen months and had prepared themselves for, in one way or another, since 1961. They were going to land on the Moon.

BACK ON THE AIR . . . MOSTLY

As both *Columbia* and *Eagle* passed behind the Moon for the last time before the landing burn—called Powered Descent Insertion or PDI—Mission Control heard Collins's voice as *Columbia* emerged from the lunar backside. "Listen, babe, everything is going just swimmingly." Collins was the only person in creation able to hear what had been occurring aboard *Eagle* since Mission Control had lost the signal forty-five minutes earlier.

Mission Control during the Apollo flights. This image is from the Apollo 10 mission.

Two minutes later *Eagle* came back into radio range, but the signal was awful. Kranz now had his first dilemma of the landing: the procedures, many of which he had written, said that they had to have both telemetry—the digital monitoring of *Eagle*'s onboard systems—and voice communications to continue the landing. In about ten minutes he would be forced make the call, go or no-go for the descent burn. It all depended on the radio signal.

The room got quieter as the controllers sweated out the communication issues. Kranz knew the rules—he had created most of them—but also knew how dangerous an abort was, and therefore had an idea of how far he was willing to

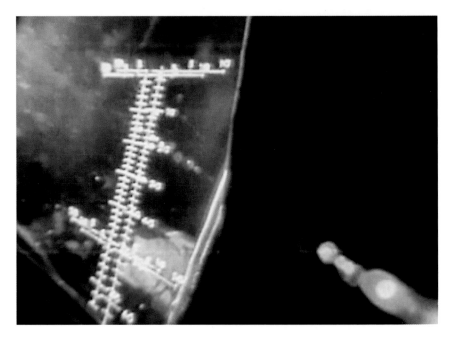

The Landing Point Designator (LDP) inscribed on the Lunar Module's windows. When Armstrong compared the landmarks below them with the LDP markers he realized that they were farther downrange than desired.

stretch those rules. But this also meant that the landing attempt was now moving a bit off the books, and that did not sit well with him. Problems had found them, Kranz later said, "like flies to a picnic lunch."[46]

As the LM crew adjusted the antennas, Kranz got bits and pieces of the information he needed—just enough to allow the landing preparation to continue. With five minutes to go, he was praying. Charlie Duke, as CAPCOM, was using the precious gaps in static, when he could actually understand what Aldrin was saying over the radio, to suggest alternative ways to adjust communications from the LM's end.

Just as Kranz had to make the go/no-go call for PDI, there was another burst of communications, and he decided it was just enough to proceed. He went "around the horn," polling each of the console positions for their recommendations.

"I'm going around the horn, make your go/no-gos based on the data you had prior to LOS—okay, we got it back, give it another few seconds . . ." The sweat was starting to build up on Kranz's collar, and many of the controllers were puffing furiously on cigarettes. "Okay, RETRO?" "Go." "FIDO?" "Go." "Guidance? "GO!" Kranz actually chuckled for a moment at the

intensity of the yelled response from his guidance officer. "TELCOM?" "Go." "GNC?" "Go." "EECOM?" "Go." "Surgeon?" "Go."

With that done, Kranz said, "CAPCOM, we are go to continue PDI." Duke passed it along. "*Eagle*, Houston. We read you now. You're go for PDI. . . . You are go to continue powered descent. You are go to continue powered descent."

About 4 minutes later, at 102 hours, 33 minutes, Armstrong pushed the "PROCEED" button on the guidance computer and said, simply, "Ignition."

Eagle was flying facedown with its windows pointing at the Moon, and as they slowed, Armstrong noticed that the landmarks he had memorized were a bit off time-wise, about two seconds ahead of when they should have appeared against the numeric indicators scribed on the window. Then the descent engine throttled down, right when it was programmed to. But the computer didn't know that it was about two seconds off, due to the added velocity they had picked up when undocking from the CM. Armstrong noted this to Buzz, and both men realized they would have some extra work to do that day—they were off-course.

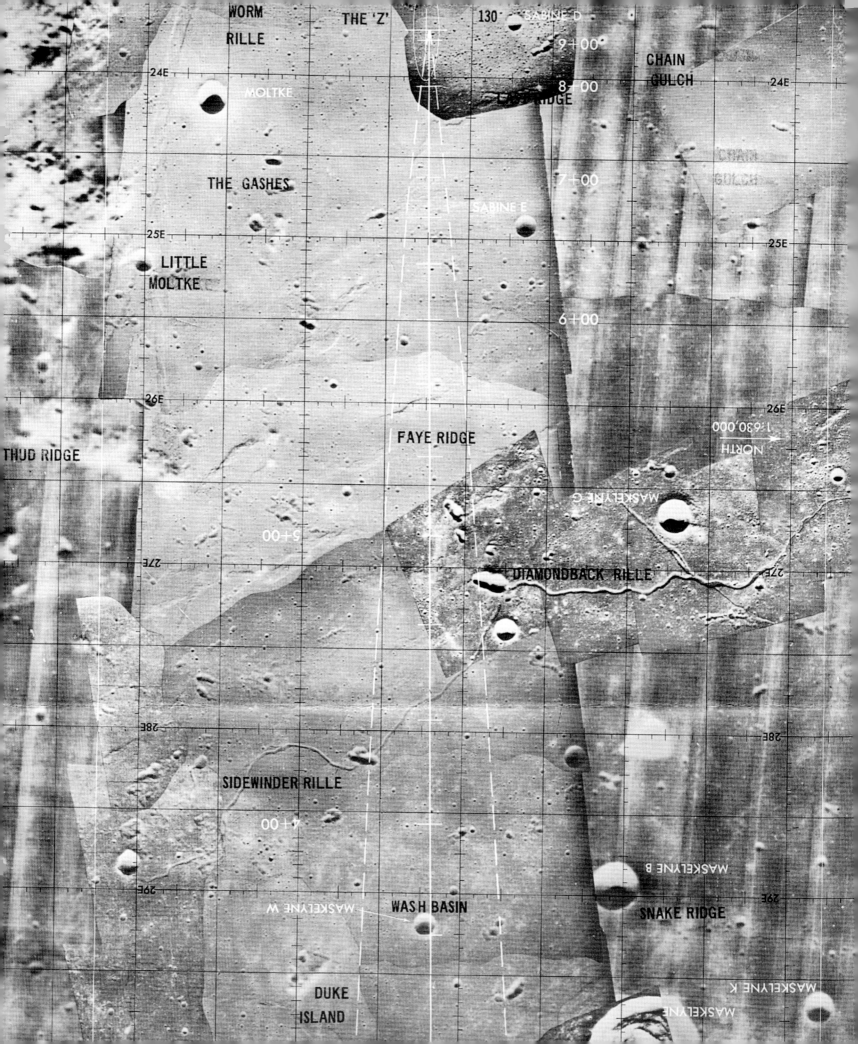

WORM
RILLE

THE 'Z'

130   SABINE D

9+00

CHAIN
GULCH

24E   24E

MOLTKE   8+00   LAST RIDGE

THE GASHES   7+00   CHAIN
                                                                        GULCH

SABINE E

25E   25E

LITTLE
MOLTKE   6+00

26E   26E

FAYE RIDGE   1:630,000

                                                        NORTH

THUD RIDGE

MASKELYNE G

5+00

27E   27E

DIAMONDBACK RILLE

28E   28E

SIDEWINDER RILLE

4+00

MASKELYNE B

29E   29E

MASKELYNE W   WASH BASIN   SNAKE RIDGE

MASKELYNE K

DUKE
ISLAND   MASKELYNE

CHAPTER 10

DOWN AND SAFE

"AFTER WE FINISH THIS SON OF A GUN, WE'RE
GONNA GO OUT AND HAVE A BEER AND SAY, 'DAMMIT,
WE REALLY DID SOMETHING!'"

—Gene Kranz, Apollo 11 flight controller

ARMSTRONG AND ALDRIN WERE descending from an altitude of about 33,000 feet (10,000 m) when the first computer alarm came in, the accusing green digits 1202 staring at them as the machine refused to give up any more information.

Armstrong said, "Program alarm," and then, "It's a 1202." Aldrin repeated it in case their comms were spotty again: "1202."

Armstrong and Aldrin went back and forth a bit about what it might be—then Armstrong said, in more urgent tones, "Give us a reading on the 1202 program alarm," to Mission Control.

On the ground, the flight controllers looked at one another, then at their clipboards if they had crib notes for computer alarms. In the back rooms, where support personnel were seated to back up the flight controllers, more notes were consulted and binders flipped through furiously. Every second could be critical.

CAPCOM Charlie Duke recalls the moment well: "I was shocked. Actually, 'stunned' is a better word. I started reaching for my guidance and

OPPOSITE: A NASA descent targeting chart prepared for the lunar landing. *Eagle* would track from the bottom to the top, with the presumed landing site at the convergence of the lines. In the actual event, *Eagle* flew well past the idealized plan.

ABOVE: Apollo astronaut Charlie Duke on the console during the Apollo 11 landing. Image from a 16mm film clip.

LEFT: The 1202 program alarm is seen here on a modern replica of the Apollo guidance computer's display.

RIGHT: Steve Bales was a young controller in Mission Control during the landing of Apollo 11. Specifically, he was the guidance officer (GUIDO), responsible for knowing the guidance system cold. His timely call, as assisted by the capable Jack Garman, allowed the landing to continue when the computer alarm sounded. He was twenty-six years old.

navigation checklist to see what [the code] was."[47] He didn't find much.

The ground loop commentary continued, mostly concerning the radar signal that had just been acquired. But the 1202 alarm was potentially urgent.

Fortunately, Steve Bales had recently prepared a cheat sheet of computer alarms after a simulation earlier that month had resulted in his calling an unnecessary abort. A simulator supervisor, called the Sim Sup for short, had tossed both the 1202 and similar 1201 computer alarms into a single simulation—and Bales had called the abort, not realizing that these alarms were important but harmless. Bales was chagrined afterward—he felt he should have known better and told Kranz he would make sure this kind of thing

did not happen again. He certainly did not know that they would be facing the same error codes on this, the first landing.

Bales also credits Jack Garman with the "save." "There's a lot going on," Bales later recalled. "And then finally the [radar] data comes in and we see the [computer alarm], and Jack is yelling—I mean, almost literally yelling, 'It's okay! It's okay, as long as it doesn't keep going on!'"[48]

Bales relayed the info to Kranz, who allowed the landing to continue.

As *Eagle* continued descending, the computer would intermittently lock up and display either the 1202 or, later, 1201 error code. What the code meant to the computer's primitive logic was, in effect,

A schematic of the guidance officer's console at which Steve Bales sat. He was responsible for a dizzying array of functions and conditions, as were all the flight controllers.

"I'm receiving too much information. If I don't reset myself, I may crash. So I will toss some of my tasks and continue with the really important ones after I sulk for a while." This is paraphrased, of course, but you get the idea.

THE 1202 CULPRIT

The culprit was the radar signal. The radar lock on the lunar surface below had come in late, and when it did, there was a mismatch between it and the rendezvous radar, which was not supposed to be feeding in a signal. As one of the engineers put it, "[It] was an obscure mismatch deep in the electronics—two signals that should have been locked together in phase were only locked together in frequency. That hardware glitch involved the rendezvous radar, which really wasn't needed during the descent to the Moon."[49]

Fortunately, the computers continued their essential tasks and the landing was able to continue with the codes popping up intermittently.

A side note: The complex and revolutionary code written for the Apollo Guidance Computer came out of MIT's Instrumentation Laboratory, which had been contracted for the work. Margaret Hamilton was the director of the Software Engineering division, and responsible for overseeing and writing much of the computer code. This was very early in digital computing, and the entire effort had to be distilled to operate in a 36-kilobyte computer that was less powerful than what would run a toaster oven today. She smartly implemented the alarm code that indicated

A WIDE PANORAMIC IMAGE assembled from a number of stills shot in sequence by Buzz Aldrin with Armstrong seen working near the Lunar Module. The small, white object sitting upright just below and to the right of the LM is the Apollo Lunar Surface Close-up Camera (ALSCC). The two rocks just below right of the camera appear to be about 2 feet (60 cm) high, while the LM in the background is 23 feet (91 cm)

tall, to give you a sense of scale on the Moon. Aldrin's shadow is seen at the bottom right of the frame. The astronauts shot many such panoramic photo series during their Moonwalk, but only with the advent of digital photo editing tools have they been improved enough for public display.

with more tasks than I should be doing at this time and I'm going to keep only the more important tasks; i.e., the ones needed for landing.' . . . The software's action, in this case, was to eliminate lower priority tasks and reestablish the more important ones. . . . If the computer hadn't recognized this problem and taken recovery action, I doubt if Apollo 11 would have been the successful Moon landing it was."[50]

With the computer emergency seemingly in hand, Kranz polled his controllers again and told Duke to relay to *Eagle* that they were go for landing. "Houston. You're go for landing, over." Aldrin replied, with remarkable calm, "Roger, understand. Go for landing." Then, almost lackadaisically, "Program alarm." It was a 1201 code, but this time, Bales jumped in immediately with "Same type, we're go." Kranz assented.

OFF-TARGET

By now, Armstrong realized he was way off target, but he was not saying much about it to Mission Control. He was intently focused on the surface outside the tiny triangular window. Aldrin was now manning the radio alone, with his eyes glued to the computer readout. He gave a steady narrative of attitude and speed. If all did not continue to go well, it could soon be his epitaph.

One of the controllers relayed a seemingly innocuous comment from his telemetry of the *Eagle*: "Attitude hold." This meant that Armstrong had switched the LM to straight horizontal flight and stopped his descent. He was looking for a landing site with the LM under manual control, flying with the hand controller. The surface below him was not only past the preferred landing zone, but also rough—the terrain was too rocky and had too many craters to assure a safe landing. The computer, however, did not know that—it only knew where it *thought* it should land. It was time to insert the humans fully back into the loop.

Meanwhile, the fuel was draining. Quickly.

They were at 400 feet (122 m) altitude. CAPCOM Charlie Duke realized what was going on and said over the loop, "I think we'd better be quiet now." What he did not say was that Deke Slayton, who was standing near him, had punched him in the shoulder and said,

Margaret Hamilton, the director of software engineering for the MIT Instrumentation Laboratory. She was responsible for most of the code that drove the Apollo guidance computer, and her insertion of the 1201 and 1202 error routines kept the computer from crashing and causing an abort. She is seen here with paper printouts of the computer code. She was later awarded the Medal of Freedom for her efforts.

the "executive overflow" problem that popped up, yet allowed the computer to keep running critical tasks. Her work later earned her the Presidential Medal of Freedom.

She later wrote of the incident, "The computer (or rather the software in it) was smart enough to recognize that it was being asked to perform more tasks than it should be performing. It then sent out an alarm, which meant to the astronaut, 'I'm overloaded

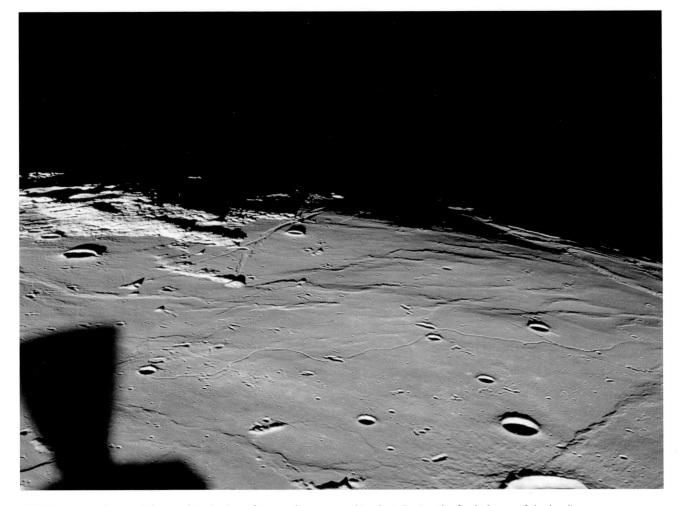

While it appeared smooth from orbit, the Sea of Tranquility was anything but. During the final phases of the landing, Armstrong struggled to find a spot "tranquil" enough to set down.

"Shut up and let 'em land!"[51] Slayton wanted Mission Control to give them less information and let the astronauts 240,000 miles (386,000 km) away do their jobs. Sheepishly, Duke agreed.

Kranz said, "Okay, the only callouts from now on will be fuel." That was now the most critical element in the complex calculus of lunar landing, and it was getting dangerously low.

Aldrin kept the numbers coming, and then said to Armstrong, "You're pegged on horizontal velocity." It was a gentle encouragement to set *Eagle* down, and soon.

"I got the shadow out there," Aldrin said. "Two fifty down at 2½. . . . Coming down nicely."

"Going to be right over that crater," Armstrong interjected.

"Two hundred feet [61 km]," Aldrin continued. Static once again crackled in the speakers; the transmissions were weakening as Armstrong searched for a smooth spot. "Five and a half down . . ." Aldrin said, referring to their rate of descent.

Still, Armstrong pressed forward. There had to be a flat spot out there somewhere.

"LOW LEVEL"

Back on the ground audio loop, a warning call: "Low level." This was indicated both on the consoles and as a light in the LM, indicating that they were almost out of fuel—it was now so low, the sensors could no longer measure it. Aldrin said, "One hundred feet (30 km) . . ." then, "Five percent," which was the remaining fuel.

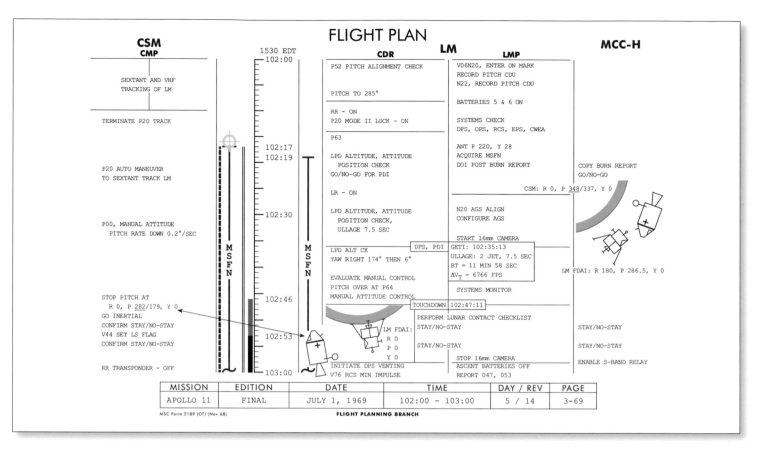

From the *Apollo 11 Flight Plan*, how the lunar landing was *supposed* to go.

One more indicator came on in the LM—"Quantity light"—a visual indicator of the same. In ninety-four seconds, they would reach a point referred to by some as the "dead man's zone," the spot in the fuel-versus-time graph at which they were supposed to abort if they could not land within twenty seconds. The issue at hand was that if they were too close to the lunar surface, nobody was sure that the LM could stage and ignite the ascent engine fast enough to avoid becoming a few acres of smashed aluminum fragments and fluttering Mylar sheets.

In addition, Armstrong still had to stop his horizontal movement and drop straight down—any sideways motion could tear off a landing leg or cause the LM to topple to the side; either would doom the crew. But he was still hovering and scooting forward, looking for a smooth place to set down the lander.

Bales remembers looking at the plot on his screen and wondering what was going on up there. He recalled that in the hundreds of simulations, Armstrong had always canceled out his forward motion early in the landing and "pretty much just came straight down.

But he wasn't. He had a forward velocity of 20 feet per second (6 mps). And, of course, that was eating up fuel."[52]

"We wanted to pick a spot that was pretty good while we still had about a hundred and fifty feet [45 km] of altitude," Armstrong explained.[53] Aldrin added that if you have a bad landing spot to the left and the right, the temptation is to "fly over," as in continue flying forward. All true, but there was not enough fuel left to fly much of *anywhere* now.

They had long ago given up trying to identify landmarks. The best maps had come from the low-altitude flyover of Apollo 10 a few months earlier, but these had been from about 8 miles (13 km) up, and flying this close, nothing matched.

"In every training run, we would have put it down by now," Kranz reflected.[54] Sometime during the final stages of the descent, the pencil he was holding in his hand snapped in two. He switched to a pen—it was sturdier.

In the back room, a controller named Bob Nance was staring at a paper chart recorder. There were lines converging that showed throttle settings and fuel level, among other things. Nance was looking at the tracks and values, then projecting ahead in his mind based on hundreds of simulated scenarios. He was generally accurate to within ten seconds or less. This is how such things were done at the dawn of the computer age—there was still a lot of seat-of-the-pants mental calculation, intuition, and even guesswork involved—both in Mission Control and above the Moon. Nance was now sweating—this was going to be a squeaker.

ONE MINUTE OF FUEL

"Sixty seconds" broke the silence in Mission Control—Charlie Duke relayed the time left until a mandated abort to the crew. Armstrong and Aldrin were still 75 feet (23 km) above the Moon. The crew was silent, concentration fixed on the job of setting down. Then Aldrin's voice came through: "Thirty feet [9 m], 2½ down. Faint shadow."

A slight ripple of relief passed through Mission Control—they just might make it, but it would be close . . .

"Four forward, four forward, drifting to the right a little," Aldrin said. Armstrong was silent.

"Thirty seconds!" said Charlie Duke. In the LM, Aldrin took his eyes off the computer readout for a moment to check the location of the "ABORT" button. One of them might be using it within moments.

The LM was still drifting sideways a bit, and Armstrong was trying to cancel that motion to bring it straight in—he had found his landing spot. He couldn't actually see it—the exhaust from the descent engine was kicking up lunar dust that had sat, undisturbed, for billions of years and was creating a great plume below. But he could see their shadow against the dust billows, and that was guidance enough. He knew he was home free. "At that altitude, running out of fuel wasn't a consideration. Because we would have let it just quit on us, probably, and let it fall on in," he said.[55]

A few seconds later, Aldrin said, "Contact light." A 5½-foot (1½-m) metal rod that extended below three of the LM's landing legs had touched the lunar surface, triggering a pale blue light on the instrument panel. Armstrong could now cut the engine and let *Eagle*

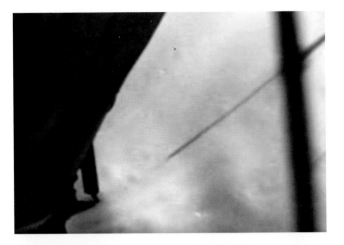

TOP: Though it's difficult to discern in this 16mm frame, filmed during the descent, dust stirred up by the descent engine made it impossible to see the surface during the final stage of the landing. The shadow of a lunar contact probe, a metal rod sticking out from three of the four landing legs, is to the right.

BOTTOM: A relieved Charlie Duke smiles immediately after touchdown.

settle onto the Moon with a muted thump—but he waited until *Eagle* set down fully.

"Shutdown." Armstrong said. Aldrin echoed, "Okay, engine stop." It was 102 hours, 45 minutes, and 43 seconds since launch.

The crew threw a few switches to shut off power to the engine and "safe" the LM. On Earth, Charlie Duke said, "We copy you down, *Eagle*."

After turning off the engine-arming circuit, a final step in making sure it was completely shut down, Armstrong said, "Houston, Tranquility Base here. The *Eagle* has landed."

Page 178   Day 5

04 06 44 53 LMP   Okay, 75 feet. And it's looking good; down a
                    half. 6 forward; light's on. 6 - 60 feet down,
                    2-1/2, 2 forward, 2 forward.

04 06 45 13 LMP   Looks good. 40 feet down, 2-1/2. Picking up
                    some dust. 30 feet, 2-1/2 down - straight down;
                    4 forward, 4 forward, drifting to the right a
                    little.

04 06 45 25 LMP   20 feet, down a half; drifting forward just a
                    little bit. Good. Okay.

04 06 45 41 CDR   SHUTDOWN.

04 06 45 42 LMP   Okay. ENGINE STOP; ACA out of DETENT.

04 06 45 43 CDR   Out of DETENT.

04 06 45 45 LMP   AUTO MODE CONTROL, both AUTO; DESCENT ENGINE
                    COMMAND OVERRIDE, OFF; ENGINE ARM, OFF; 413 is in.

04 06 45 52 CDR   ENGINE ARM is OFF.

04 06 45 58 CDR   Houston - Tranquility Base here. THE EAGLE HAS
                    LANDED.

04 06 46 14 CDR   Thank you.

04 06 46 17 CDR   Okay. Let's go on. Okay, we're going to be
                    busy for a minute.

04 06 46 23 LMP   Alright, MASTER ARM, ON. Take care of the descent
                    vent.

04 06 46 25 CDR   MASTER ARM coming OFF.

04 06 46 27 LMP   I'll get the pressure vent.

04 06 46 28 CDR   Okay.

04 06 46 36 LMP   Very smooth touchdown.

04 06 46 49 CDR   I didn't hear that vent going - -

04 06 46 51 LMP   ... oxidizer.

ON THE MOON AT LAST

A relieved, and temporarily flustered, Duke responded, "Roger, Twan . . . [correcting himself] Tranquility. We copy you on the ground. You got a bunch of guys about to turn blue. We're breathing again. Thanks a lot."

Armstrong replied, "Okay, we're going to be busy for a minute." He and Buzz now had to reset the LM for the possibility of an immediate emergency liftoff, in case something went wrong.

While Armstrong and Aldrin smiled, shook hands, and completed their post-touchdown checklist, there was a moment of well-earned celebration in Mission Control. In the glass-enclosed visitor's area, it was pandemonium. At the control consoles, there were some loud congratulations, a lot of backslapping and hand-shaking, and a few silently shed tears. Their decade of endless training and simulations had been rewarded with a successful landing. But they were not home safe yet.

After a brief moment of emotion, Kranz, ever the marine, snapped back to it. First he asked for a stay/no-stay decision from his team. Then, he barked, "All right, keep the chatter down in this room!" The controllers quieted. All systems looked good for Armstrong and Aldrin's short stay on the Moon.

The relief lasted for a relatively tranquil fifteen minutes. Then Kranz got a call from a controller who said, "Flight, the descent-engine helium-tank pressure is rising rapidly. The back room expects the burst disk to rupture. We want the crew to vent the system."

This was not good. Something was causing pressure to build up in a fuel line in the descent engine. Tom Kelly and his people from Grumman and TRW (builders of the descent engine) were at Mission Control for the landing and looked temporarily stricken, then they began discussing the issue. Kranz wanted a answer immediately.

The best guess was that an ice plug had formed in a fuel line in the descent stage—the cold of the surface had overridden the heat inside the lower half of the LM, causing some fuel to freeze. If the pressure behind that ice plug kept rising, the best-case scenario was that a "burst disk"—a pressure-relief system—might rupture. That wasn't too bad; the descent stage had done its job, never to be used again. The worst case, however, was not as nice—the lower half of the LM could potentially explode, killing the crew.

Should Kranz call an abort and have the crew return to orbit immediately? Or should he follow the advice of the engineers and "burp" the engine to see if that relieved the pressure? That plan had its own risks—opening the valves for even a moment would result in a brief combustion, and *Eagle* could end up "hopping" or even tipping over. Fortunately, while all this was being considered, the ice plug in the fuel line melted, and pressures dropped. Problem solved. . . . Now on to whatever was next.

Below the astronauts, in the ticking and cooling descent stage, less than one-minute's worth of fuel settled into the tanks. When they had landed, this was all that stood between them and a mandatory abort. It was a thin margin.

OPPOSITE: NASA's official transcript of the final moments of the landing of Apollo 11. Armstrong's confirmation of touchdown from the surface of the Moon is enthusiastically circled with an arrow pointing to it.

ABOVE: A row of flight controllers is quietly jubilant immediately after touchdown. 16mm film frame.

CHAPTER 11

MOONWALKING

"THAT'S ONE SMALL STEP FOR MAN . . . ONE GIANT LEAP FOR MANKIND."

—Neil Armstrong, Apollo 11 astronaut

IT'S UNCLEAR IF THE WORDS ABOVE are what Armstrong actually heard inside his helmet. He swore to the end of his days that he had intended to say, "That's one small step for *a* man," because otherwise, he felt it didn't make any sense. Historians and technicians have listened to the recordings from that day again and again, more recently using advanced software to enhance the audio and even using voice pattern-recognition routines in an attempt to identify any missing words. It has become an almost quixotic quest that has not yielded a definitive answer. But we know what he wanted to say, and that's good enough.

After the crew prepared for a possible emergency liftoff, which involved setting switches and loading new information into the guidance system, they had a bit of time for personal tasks. Both marveled at the view outside the window—*Eagle* had come to rest on a broad, flat plain, pockmarked with craters of varying sizes and with rocks and boulders everywhere. In the distance were some low ridges, about 20 to 30 feet (6 to 9 m) high. Armstrong was glad he was able to avoid all the obstructions—if the

LM had landed at too much of a tilt, their departure could have been impaired.

There were no immediately identifiable features, and Armstrong said wryly, "The guys that said that we wouldn't be able to tell precisely where we were are the winners today," eliciting a few muted chuckles in Mission Control.[56]

Aldrin, with a gift of description that would become apparent during the Moonwalk, said: "It looks like a collection of just about every variety of shape, angularity, granularity, about every variety of rock you could find. The color . . . varies pretty much depending on how you're looking relative to the zero-phase point. There doesn't appear to be too much of a general color at all. However, it looks as though some of the rocks and boulders, of which there are quite a few in the near area . . . it looks as though they're going to have some interesting colors to them."[57] The geologists in the back room in Houston listened raptly.

OPPOSITE: Aldrin prepares to take the final step down the ladder, as photographed by Armstrong. Most of the images of the Moonwalk were taken by Armstrong, since he carried the camera.

ABOVE: The first image taken by humans on another world—Armstrong took a series of images, assembled here, out the window of the Lunar Module shortly after landing. The shallow crater to the right is about 36 feet (11 m) across. Had one of *Eagle*'s footpads landed inside the crater, it could have been enough to ruin their chances of a safe liftoff from the Moon.

RIGHT: As soon as the landing procedures were complete, Kranz turned over Mission Control to another flight director. Kranz's job was done for the day, save for making notes about the landing for future reference.

Armstrong and Aldrin had some technical house-keeping to do. Next on the timeline was their "rest period," a scheduled nap that neither of the pair was particularly interested in, but it had been inserted into the flight plan at the insistence of the doctors. While they needed to decompress a bit, sleep was unlikely.

This gave them a bit of time for personal reflection, and for Aldrin, part of that would be a moment of engaging his faith.

Religion and spaceflight have not always been comfortable together, though many of the astronauts were regular churchgoers. During the flight of Apollo 8, the crew read from Genesis in the Bible while in orbit around the Moon on Christmas Eve. While it was a brilliant bit of heartfelt PR and was generally well received, some felt that it was inappropriate, and at

least one person sued NASA. Madalyn Murray O'Hair, the founder of the group American Atheists, alleged infringement of First Amendment rights. The Supreme Court dismissed the suit, but NASA was now gun-shy over religious expression in space.

Aldrin, however, was determined to have a brief ceremony of personal observance, and he pulled a tiny pack out of a pouch that contained a miniature chalice and some Communion wafers. He poured a tiny bit of wine into the little cup and said, "This is the LM pilot. I'd like to take this opportunity to ask every person listening in, whoever and wherever they may be, to pause for a moment and contemplate the events of the past few hours and to give thanks in his or her own way. Over." And with that, he performed his miniaturized ceremony, with Armstrong looking on. Aldrin read words from the Book of John silently, so as to not stir up any more debate.

While Armstrong had previously approved the doctor-mandated rest period after landing and before they would exit the LM, both he and Aldrin were ready to go and explore—sleep be damned. They got permission from Mission Control to begin early and ate some dinner in preparation for their excursion.

After the meal, the two Moonmen—the first beings in history that could rightfully claim the title—started the long process of donning their EVA suits and life-support backpacks, checking each item as they went, then cross-checking each other's. The backpacks,

called Portable Life Support Systems, or PLSS, packs, were another of the Apollo program's miracles of miniaturization. Each could support a human on the lunar surface for hours using oxygen, carbon dioxide scrubbers, and a water supply for both drinking and keeping the astronauts at a comfortable temperature via circulation through the pressure suit. Outside, their suits would be exposed to an approximately 500-degree-Fahrenheit (260°C) differential between the lit side and the shadowed side, and a system of plastic tubes was woven into the suit undergarment through which the water supply would be circulated to equalize their temperature.

Helmets and gloves went on last, and then it was time to depressurize the cabin and open the door. This proved more difficult than anticipated. Despite venting the atmosphere from the LM into the lunar vacuum, enough air remained that the inward-opening hatch

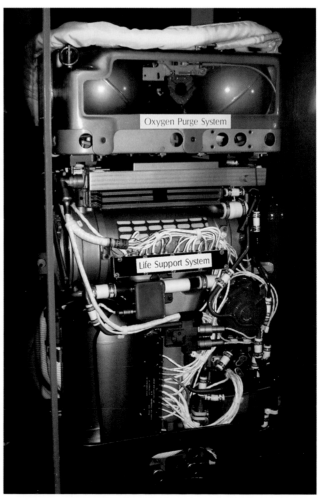

ABOVE: This view out the hatch of the Lunar Module during training gives you an idea of what the astronauts might have seen when they opened the hatch on the Moon—minus the shoes of the technician outside. At center is Charlie Duke's Portable Life Support System unit (from Apollo 16), and to left and right are the helmet-stowage bags.

RIGHT: One of the Apollo Portable Life Support System, or PLSS, backpacks. They were a magnificent invention of the Apollo program, and none ever failed on the lunar surface.

ONE GIANT LEAP

Armstrong got down on all fours, with his feet sticking out of the hatch. Somewhere in that process he apparently snapped off a protruding plastic switch on one of the LM's control panels—as luck would have it, the arming switch for the ascent engine. This was only discovered later when going over checklists in preparation to depart the Moon.

It was a cramped space, and everything had to be done with care. "The LM structure was so thin one of us could have taken a pencil and jammed it through the side of the ship," Armstrong recalled.[58] Care was the watchword, but it was not easy. "We felt like two fullbacks trying to change positions inside a Cub Scout pup tent," he added.

Aldrin, clad in his own Moonsuit, talked Armstrong through the hatch. "All right. Move . . . to your . . . roll to the left. Okay. Now you're clear. You're lined up on the platform. Put your left foot to the right a little bit. Okay. That's good. Roll left. Good."[59]

Armstrong crawled onto a platform—called the "front porch"—that sat on the front leg of the LM, grasped the handrails on either side, and moved toward the edge of the porch. Before descending, he reached over and tugged on a handle near the left side of the platform. This was a manual release for a hinged panel on the side of the LM that, when released, swung down to reveal stowed equipment and, more important at the moment, a TV camera.

A ghostly image sputtered to life on the screens at Mission Control. Although it was in black and white

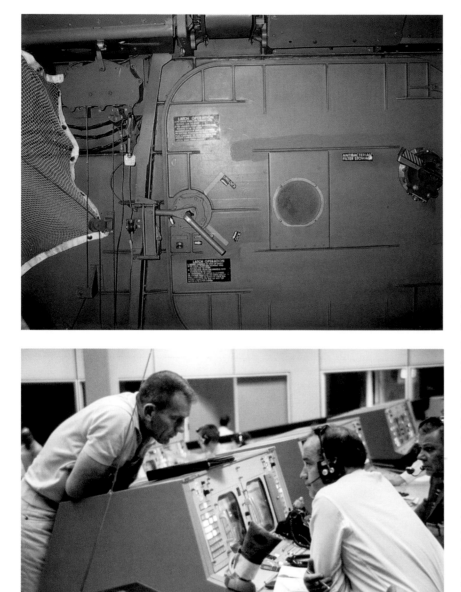

TOP: The hatch of the Lunar Module was difficult to open when it was time to exit the spacecraft. Aldrin had to flex the upper left edge to release the remaining air inside before it would open.

BOTTOM: Deke Slayton, head of the astronaut office, confers with Paul Haney, the "voice of Mission Control" during the Moonwalk.

refused to budge. There was still a tenth of a pound (45 g) of pressure pushing against it. They struggled briefly, and then Aldrin reached to the upper edge of the hatch and tugged. It flexed enough to allow the residual air to escape—the hatch was that thin—then swung open. The Moon, bright gray and tan, beckoned.

Although there is no picture of Armstrong descending the ladder of the Lunar Module on Apollo 11, this photo of Pete Conrad departing the LM on Apollo 12 shows what Armstrong must have looked like right before his historic first steps on the Moon.

TOP: Armstrong preparing to take his first step off the Lunar Module footpad and onto the surface of the Moon. The quality of the TV image was low-grade by today's standards.

RIGHT: Before taking his first step, Armstrong practiced jumping back up onto the first ladder rung to make sure he could. The lander's leg had not compressed as far as expected, and the bottom rung of the ladder was about 32 inches (81 cm) above the footpad.

and fuzzy, it was glorious. There was a man, crouched on the front of a machine that had just landed on the Moon! Unfortunately, the image was also upside down. It took a few moments to get it inverted so that it appeared to be properly oriented—it was just the throw of a switch at the tracking station. Now that Armstrong had done his part, he was intent on getting down the ladder.

There had been concerns about how high above the ground the lower end of the ladder might be. Inside the LM's legs were simple, collapsible mechanisms

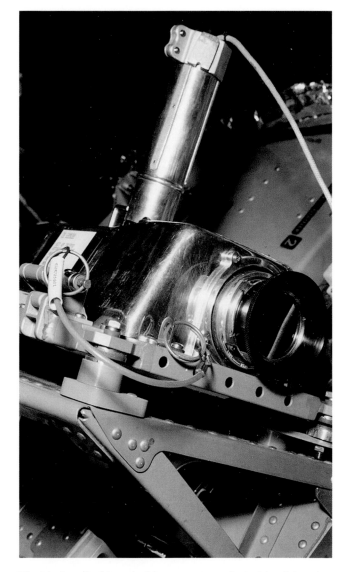

The black-and-white television camera on the inside of the flip-down panel on the side of the Lunar Module. Armstrong yanked a handle before descending the ladder to release the camera.

that crumpled to absorb the shock of landing. How far these would compress was an educated guess based on testing, but in the final event, the amount of compression depended on the speed the LM was going when it touched down on the Moon. Armstrong had kept the descent engine running all the way to the lunar surface (subsequent flights turned off the rocket before touchdown), so the legs had not compressed far.

Armstrong, cognizant of this, got his boots onto the lower rung, then dropped down so that he was standing on the LM's footpad, a wide dish on the bottom of the landing leg. Without releasing his grip on the handrails to either side of the ladder, he pulled himself back up to make sure his feet could reach the bottom rung again.

"Okay. I just checked getting back up to that first step, Buzz," he said. "The strut isn't collapsed too far, but it's adequate to get back up."

With that accomplished, he lowered himself so that he was standing once again on the footpad. This was the big moment.

As he prepared to set foot on the Moon, Armstrong made sure to note the condition and position of the footpad to alleviate any concerns at Mission Control about the stability of the LM—some had worried that it might shift after landing. There were still many unknowns about the surface dynamics of the Moon. "I'm at the foot of the ladder. The LM footpads are only depressed in the surface about 1 or 2 inches [2.5 to 5 cm], although the surface appears to be very, very fine grained, as you get close to it. It's almost like a powder. [The] ground mass is very fine," he said. Then, a second later, "I'm going to step off the LM now."

"MAGNIFICENT DESOLATION"

All eyes in Mission Control, and across the world where people were able to watch TV, were glued to the screen. Elsewhere, people sat fixated on their radios, their imaginations reaching across the 240,000 miles (386,000 km) of empty void to the Moon.

Armstrong stepped. "That's one small step for man . . . one giant leap for mankind."

An artist's impression of how Armstrong's

The first picture Armstrong took during his EVA. The white cloth bag is a cover from the Mesa, the stowage bay on the side of the Lunar Module.

Reporters around the world scribbled on notepads and rushed to payphones to call in the "first words from the Moon." Others sat transfixed, writing occasional notes.

"The surface is fine and powdery," Armstrong said. "I can kick it up loosely with my toe. It does adhere in fine layers, like powdered charcoal, to the sole and sides of my boots. I only go in a small fraction of an inch, maybe an eighth of an inch [3 mm], but I can see the footprints of my boots and the treads in the fine, sandy particles."

Next Aldrin passed a camera down to Armstrong so that he could take a photo panorama of the landing site. They used a clothesline-like contraption to do so— it was officially named the Lunar Equipment Conveyor, or LEC, but the astronauts referred to it simply as "the clothesline." With camera in hand, Armstrong took a full panorama of the region surrounding him.

At Mission Control, there was some confusion. The first goal upon reaching the surface was supposed to be the grabbing of what they called the *contingency*

RIGHT: Armstrong at the ladder of the Lunar Module. This image, snapped from a monitor at the Australian receiving station, was one of the best taken of the TV transmission. Both the Goldstone radio dish in California and the Honeysuckle Creek station in Australia were receiving the transmission, but the Australian version was vastly clearer.

BELOW: Armstrong taking photographs just after descending the ladder to the lunar surface, while Aldrin waited inside the Lunar Module. This image is from a 16mm movie frame taken from the LM's window. It is unique in that it is a rarely seen image of Armstrong—there were few taken of him by Aldrin—and that his gold visor is up, making his face visible.

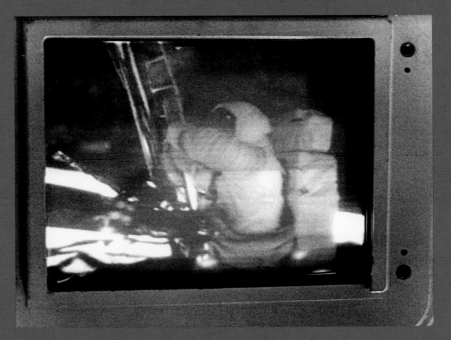

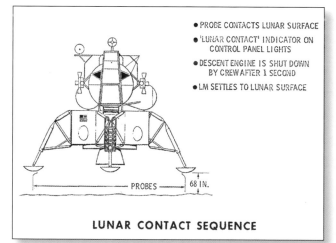

- PROBE CONTACTS LUNAR SURFACE
- 'LUNAR CONTACT' INDICATOR ON CONTROL PANEL LIGHTS
- DESCENT ENGINE IS SHUT DOWN BY CREW AFTER 1 SECOND
- LM SETTLES TO LUNAR SURFACE

- PROBES -- 68 IN.

LUNAR CONTACT SEQUENCE

TOP: Aldrin almost to the bottom rung. He would soon turn and survey the lunar desolation.

ABOVE: The final configuration of the Lunar Module. It did indeed look like it had the mumps.

OPPOSITE: Armstrong practicing ascending the Lunar Module's ladder, testing to see if his suit is flexible enough to reach the first rung.

sample, a bit of lunar soil near where Armstrong stood—gathered quickly in case they had to make an emergency departure for some reason. It would be a shame to go all the way to the Moon, declare an emergency, and depart without at least a small sample of what they had worked so hard to reach.

But Armstrong had his reasons for taking the photos first—the LM was in deep shadow on this side, and he wanted to shoot the photos in that light. Not knowing this, Houston reminded him to get the sample as soon as possible. "Roger," Armstrong replied, "I'm going to get to that just as soon as I finish these . . . picture series."

Within minutes he had assembled a handle with a scoop and grabbed some soil, depositing it into a small bag that he then stuffed into a pocket. The handle was needed due to the fact that the bulky suits limited the astronauts' movement, and bending over to grab things from the surface was not possible.

"I'll try to get a rock in here. Just a couple," he said.

A few minutes later, Aldrin made his way down the ladder. When he stepped off the footpad and took in the view, he was stunned.

"Beautiful view!" he said.

"Isn't that something?" Armstrong replied. "Magnificent sight out here."

Aldrin paused, then made what is perhaps the most poignant remark ever about the bleak splendor of the lunar surface. It came out as a simple statement, with no added drama—none was needed. "Magnificent desolation," he said, almost dreamily.

An estimated 600 million people watched on Earth, almost unbelieving. Two Americans were exploring off the edge of any map, daring to do what had seemed impossible.

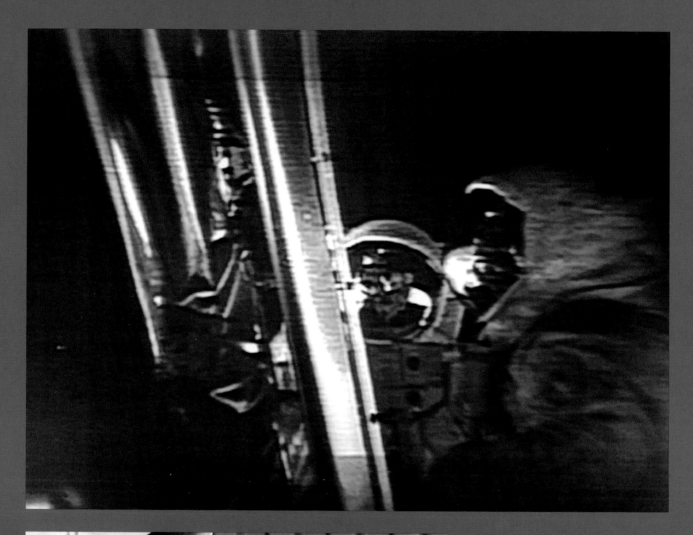

TOP: Armstrong and Aldrin read the dedication plaque on the Lunar module. It ended with the words, "We came in peace for all mankind."

LEFT: The Apollo 11 plaque and protective cover before they were affixed to the front leg of the Lunar Module.

LEFT: The view from Mission Control as Armstrong and Aldrin explored the lunar surface.

ABOVE: Armstrong and Aldrin erect the US flag on the Moon. This image is from a 16mm movie frame, captured by a camera running inside the Lunar Module. This is the only photographic source that shows both astronauts working on the Moon.

GETTING TO WORK

The astronauts had just over two hours to complete this first reconnaissance of the Moon. It was not long enough by a factor of ten. Later expeditions would have two, then three Moonwalks scheduled and also carry an electric Moon cart—the Lunar Roving Vehicle—to extend their explorations. But this first mission had been deliberately kept simple, and time was at a premium.

Armstrong and Aldrin got to work immediately. They spent a little bit of time getting acclimated to moving around on the Moon. "While the gravity is less, and you weigh less, you still have the same mass," Aldrin later said.[60] "You had to be careful when moving or hopping." Eventually, through some experimentation, Aldrin arrived at what some have termed the "lunar bunny hop" as an efficient way of getting around.

Aldrin relayed his observations to Mission Control. "Got to be careful that you are leaning in the direction you want to go, otherwise you [appear] slightly inebriated. . . . In other words, you have to cross your foot over to stay underneath where your center of mass is," he said.

Armstrong moved to the ladder of the LM for a brief ceremony, the unveiling of a plaque on the front leg of the lander. Aldrin joined him, and Armstrong read the inscription.

"For those who haven't read the plaque, we'll read the plaque that's on the front landing gear of this LM.

ABOVE: Television image of Armstrong and Aldrin erecting the flag on the Moon.

OPPOSITE: Aldrin sets up the solar-wind experiment, basically a sheet of aluminum foil that would collect radiation particles from the sun for the two-plus hours the astronauts would be on the Moon's surface.

First there's two hemispheres, one showing each of the two hemispheres of the Earth. Underneath it says, 'Here men from the planet Earth first set foot upon the Moon, July 1969 A.D. We came in peace for all mankind.' It has the crew members' signatures and the signature of the president of the United States."

The two men then set themselves to running the TV camera, now mounted on a tripod, out to a location from which it would see their activities in wide angle. Armstrong, true to his inquisitive nature, got momentarily distracted by what he thought was glass at the bottom of a small crater, but Aldrin prompted him to finish placing the camera as he fed Armstrong cable. Armstrong would later try to find the glass he thought he had spotted, but without success.

This complete, Aldrin moved on to deploy an experiment called the Solar Wind Collector that used a sheet of thin aluminum foil suspended by a frame. This would sit in one spot as they worked on the Moon, collecting ions streaming from the sun. Without Earth's atmosphere and well beyond our planet's protective magnetosphere, the experiment would be the first physical indication of exposure of materials to raw solar energy (along with the space suits and the hull of the CM, but the latter would be scalded by reentry).

Minutes later, they were planting the US flag on the lunar surface. Each moment was invaluable, and they had to work fast in this new and foreign environment. The flag was made of nylon and mounted on a crossbar that hinged from a metal flagpole. Since there is no atmosphere on the Moon, the only way the flag could "wave" was if it was held out horizontally with a metal rod. This horizontal bar did not extend fully

as designed, but few at home noticed. The astronauts pounded it as far into the soil as they could, but the flagpole was still a bit wobbly. It would later fall to the ground when struck by the blast of the departing lunar module, but it stood up during the Moonwalk, long enough to garner salutes from the astronauts.

They worked a bit more, and then there was a call from Mission Control, asking them to move back over to the flag. "Neil and Buzz," said the CAPCOM, "the president of the United States is in his office now and would like to say a few words to you. Over."

PHONE CALL ON THE MOON

Armstrong responded, "That would be an honor."

President Nixon came on the line. In the television broadcast, his image was split-screened with that of the Moonwalkers so that audiences could see them all—Nixon on the phone from the White House, talking to astronauts on the Moon. It was all at once surreal and patriotic.

Nixon began in characteristically friendly tones. "Hello, Neil and Buzz. I'm talking to you by telephone from the Oval Room at the White House, and this certainly has to be the most historic telephone call ever made. I just can't tell you how proud we all are of what

you have done. For every American, this has to be the proudest day of our lives. And for people all over the world, I am sure they, too, join with Americans in recognizing what an immense feat this is. Because of what you have done, the heavens have become a part of man's world. And as you talk to us from the Sea of Tranquility, it inspires us to redouble our efforts to bring peace and tranquility to Earth. For one priceless moment in the whole history of man, all the people on this Earth are truly one; one in their pride in what you have done, and one in our prayers that you will return safely to Earth."

There was a long pause, and then Armstrong responded, "Thank you, Mr. President. It's a great honor and privilege for us to be here representing not only the United States but men of peace of all nations, and with interests and the curiosity and with the vision for the future. It's an honor for us to be able to participate here today." Once more, men, women, and children all over the globe were briefly united in wonder.

And then they went immediately back to work. The call lasted about one minute and fifteen seconds and cost millions of dollars in "Moon-time." One wonders if it would had made more sense for the president to call when they were back inside the LM, but that would not have offered the same photo opportunity as American astronauts standing next to the flag.

In just over a year, Nixon would oversee the cancellation of the flights of Apollo 18, 19, and 20, for which hardware had already been built. The rockets and spacecraft were either sent to museums or repurposed for missions such as Skylab and Apollo's swan song, the Apollo-Soyuz Test Project. But for now, NASA was experiencing its finest hour.

After the call, Armstrong started making trips away from the LM to collect more rocks and soil. Called the *bulk sample*, it was to be gathered as far from the LM as reasonably possible to avoid

President Richard Nixon speaking to the Apollo 11 astronauts while they take a break from their lunar exploration. John F. Kennedy's legacy had become Nixon's reality.

Aldrin circled the Lunar Module, taking photos to document its condition. The spacecraft had weathered the descent to the lunar surface well.

contamination from the rocket plume and disturbance from the exhaust. By Armstrong's estimates, he made well over a dozen short treks to find representative samples and return them to the LM. As he said in a mission debriefing, "I probably made twenty trips back and forth from sunlight to shade. I took a lot longer, but by doing it that way, I was able to pick up both a hard rock and ground mass [soil] in almost every scoopful. . . . This was at the cost of probably double the amount of time that we normally would take for the bulk sample."

Lee Silver, a professor of geology from the California Institute of Technology who was recruited to train the astronauts about geological sample collecting, was

Aldrin setting up the Early Apollo Scientific Experiment Package.

impressed. "What Neil did in the shortest period of time that anybody [had] was so brilliant from this point of view of providing the materials to the scientists, that nobody can claim to have exceeded it in production per minute. He was really outstanding." Part of what impressed Silver was that Armstrong was actually breaking mission rules to collect the best samples possible. NASA had given him "a very strict protocol," Silver said, "which said, 'You will never leave the field of the [TV] camera.' Neil Armstrong recognized that just beyond the field of the camera was a rim of craters covered with rocks and dust, which had been excavated from a little deeper than

everywhere else, and he had a very special box for bringing back good samples with a special seal on it, and for about seven or eight minutes, you couldn't see Neil."[61]

In the meantime, Aldrin circled the LM, taking photos and inspecting it for any damage suffered during the landing. Other than some rumpled metal on the plume deflectors of the maneuvering thrusters, everything appeared to be shipshape. "I don't note any abnormalities in the LM," he said. Looking at the maneuvering thrusters, called *quads*, he noted, "The quads seem to be in good shape. The primary and secondary struts are in good shape. Antennas are all

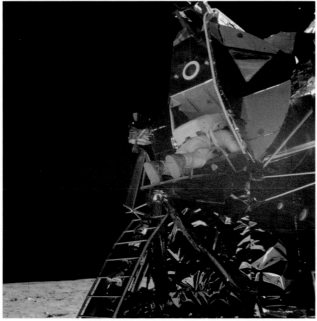

ABOVE: The Laser Ranging Retro Reflector (LRRR) set up by Apollo 11. It still works.

LEFT: A composite image of the Apollo 11 Lunar Module on the Moon with Aldrin preparing to descend the ladder.

in place. There's no evidence of problems underneath the LM due to either engine exhaust or drainage of any kind." The LM had turned out to be a very sturdy spacecraft. Now all it had to do was get them back to the CM so they could go home.

The next major task was to set up an experimental package designed to remain on the Moon. It was called Early Apollo Scientific Experiment Package, or EASEP. This was the forerunner of a more elaborate experimental package that would be flown on later Apollo missions—the latter would use a nuclear power supply, but EASEP had only two solar panels to power it. It

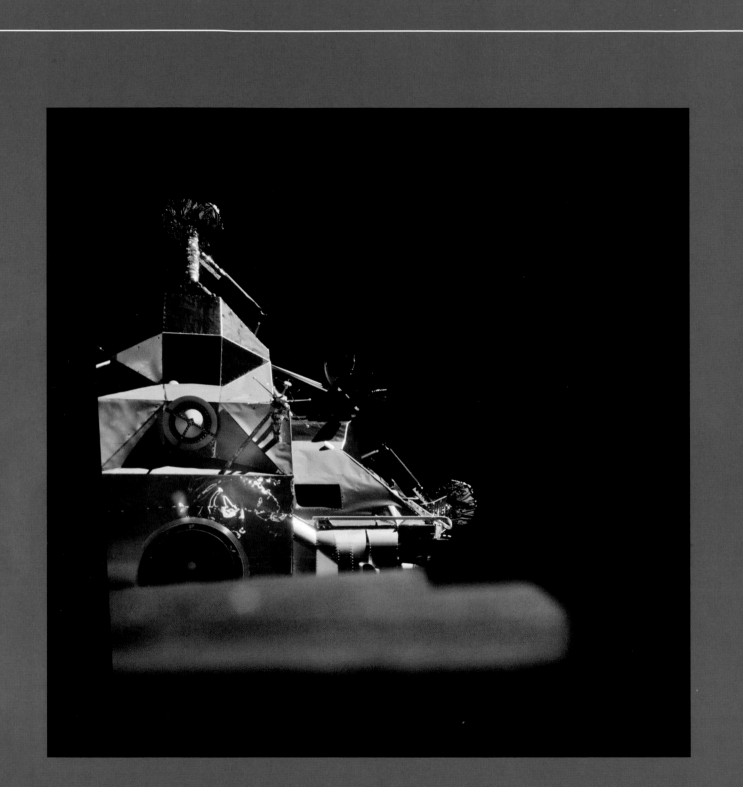

THE LUNAR MODULE IS seen undocking through the window of the Command Module in this composite image. The top hatch of the LM is seen to the lower left, and at middle left is the docking target that the CM pilot aimed for during maneuvers. The spindly looking protrusion to the right of the docking target is the VHF (very high frequency) antenna that helped the astronauts communicate with the CM, critical during docking and their descent to the Moon, when the high-gain antenna was being troublesome. The left landing leg (from the pilot's perspective) is seen to the top, and the black-and-gray panel immediately below is a fuel tank for the Ascent Propulsion System (APS). The tunnellike structure just visible on the lower right (above the out-of-focus part) is the top of the forward crew cabin above the hatch they used to exit to the lunar surface.

SHOWN HERE IS A modern, digital assembly of a 360-degree panorama shot of the Apollo 11 landing site. While relatively smooth in comparison to many later Apollo landing sites, the ruggedness that made this first landing a challenge is clear in this shot—engineers were concerned that a large rock, a small crater, or even just a hummock more than a foot or two high could have made it difficult for the Lunar Module's upper stage to perform a proper liftoff. On this first mission, caution was the watchword.

consisted of a seismometer to measure "Moonquakes," a lunar dust detector, and a radio for communicating with Earth. Another experiment was also deployed nearby, known as the Laser Ranging Retro Reflector, or LRRR. This was a box of mirrors ingeniously designed so that when a laser beam shined from Earth struck it, the return beam would follow the same exact path back. With this, astronomers were able to determine the distance between the Earth and Moon to within inches. It is still there and still operational. The EASEP operated for about a month.

FRUSTRATING MOMENTS

While Aldrin finished setting up the EASEP, Armstrong went off to inspect and photograph some large boulders nearby. He noted that the surface of the rocks had what later Moonwalkers referred to as "zap pits," mini-craters formed when micrometeorites smacked into the rocks.

Aldrin was struggling to get the EASEP as level as possible—it had a small cup with a clear cover and a metal bead that was supposed to roll to the center when it was flat. Aldrin swore that the cup had somehow

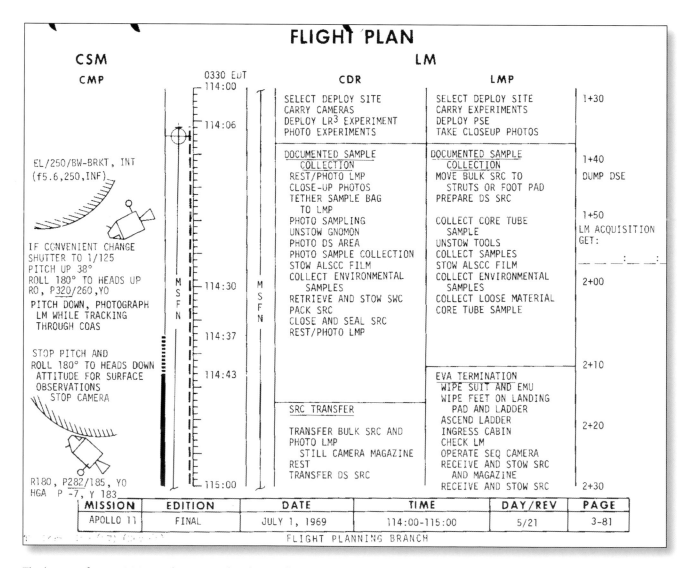

The lunar surface activities as documented in the *Apollo 11 Flight Plan*. Note that to the lower right, the astronauts are instructed to reenter the Lunar Module at two hours plus twenty minutes—Mission Control ended up giving them extra time to explore.

In this fuzzy TV image, Aldrin can be seen running toward the left of the Lunar Module.

inverted itself sometime after launch. He was wasting time trying to level the unit with little success; the little ball rolled defiantly around the outer rim. Frustrated, he radioed Houston.

"Houston, I don't think there's any hope for using this leveling device to come up with an accurate level. It looks to me as though the cup here, that the BB is in, is now convex instead of concave. Over." The CAPCOM responded, "Roger, 11. Press on. If you think it looks level by eyeball, go ahead."

Aldrin then deployed the EASEP's solar panels and prepared to move on to the next item on the checklist. Right about then, the two were reminded of the brevity of their stay—the CAPCOM told them that they had been out on the surface for two hours and twelve minutes, just a few minutes shy of the planned duration of the EVA. Houston softened the blow by adding that they were allowing an extra fifteen minutes beyond the

scheduled time in an effort to accomplish as much as possible. "Okay. That sounds fine," said Armstrong.

Houston asked for a photo of the troublesome leveling device on the EASEP, and since Armstrong was closer and had the only camera between them, he volunteered to take it.

"Oh, shoot! Would you believe the ball is right in the middle now?" he exclaimed when he got there. A slightly miffed Aldrin responded, "Wonderful. Take a picture before it moves!"

Armstrong still had more rock and soil gathering, called the *documented sample*, on his checklist. For this, he would photograph and label each item he picked up to provide context for the geologists once it was returned to Earth. At this point the astronauts were no longer walking or slowly hopping across the Moon, they were running to accomplish their tasks. Lunar soil sprayed from their boots as they ran to and fro to finish on time.

Meanwhile, Aldrin grabbed some tools from the LM and went off to collect some core samples. This involved pounding a hollow tube into the surface and extracting it, which would pull up some of the deeper history of the Moon. But it was harder than they had anticipated; Aldrin was struggling to get it as deep as they wanted. "I hope you're watching how hard I have to hit this into the ground, to the tune of about 5 inches [13 cm], Houston," he said, continuing to whack the sample tube. Toward the end the exercise, he was raising his geology hammer to the height of his helmet to

get enough force to drive the tube deeper.

"I found that wasn't doing much at all in the way of making it penetrate further," Aldrin said later. "I started beating on it harder and harder, and I managed to get it into the ground maybe 2 inches [5 cm] more. I found that, when I would hit it as hard as I could and let my hand that was steadying the tube release it, the tube appeared as though it were going to fall over. It didn't stay where it had been pounded in. This made it harder, because you couldn't back off and really let it have it. . . . I was hammering it in about as

Aldrin struggles to pound the core sample tube deep enough to collect soil from well beneath the lunar surface.

hard as I felt I could safely do it."[62] Obtaining core samples continued to vex later Apollo crews on the Moon, even with the addition of an electric-drill drive.

OVER ALL TOO SOON

Then, just as they were getting the hang of working on the Moon, a message from Houston reminded them once again that time was short. "Neil, this is Houston. We'd like you all to get two core tubes and the solar-wind experiment; two core tubes and the solar wind. Over." It was an instruction to start gathering things that had to come home with them. Armstrong responded with a curt "Roger."

The CAPCOM followed up with, "Buzz, this is Houston. You have approximately three minutes until you must commence your EVA termination activities. Over." Aldrin acknowledged.

Armstrong realized how little they had accomplished given what kinds of exploration was possible on the Moon. "There was just far too little time to do the variety of things that we would have liked to have done. [There were] rocks in a boulder field out Buzz's window that were 3 and 4 feet [1 m] in size—very likely pieces of lunar bedrock, and it would have been very interesting to go over and get some samples of those. We have the problem of a five-year-old boy in a candy store. There are just too many interesting things to do."[63]

Reluctantly, the astronauts prepared to wrap up the first exploration of the Moon. They gathered the boxes of rocks and soil that had been collected, dusted off their suits as best they could, and then Aldrin headed up the ladder.

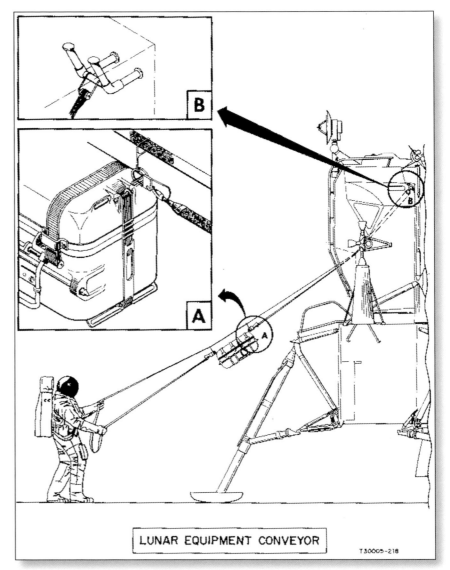

LUNAR EQUIPMENT CONVEYOR

T30005-218

A schematic of the Lunar Equipment Conveyer, which the astronauts called the "clothesline." It was fussier to use on the Moon than it had been in practice sessions on the Earth.

Meanwhile, Mike Collins, who had been orbiting overhead and listening to what parts of the Moonwalk he could through a relay from Houston, performed his own experiments and photographed the surface below. He had tried for hours to locate the LM with the onboard navigational telescope without success and so had settled into a busy routine of planned activities. He was pleasantly distracted with these when Mission Control radioed up that the first laser pulse sent to the Laser Ranging Retro Reflector had been successful. Collins was thrilled. Historians have since determined that the actual first accurate return was obtained a month later, in August, but in any event, the news kept Collins in the loop.

Once Aldrin was inside the LM, Armstrong again used the clothesline to convey the lunar samples up to the hatch. It was challenging work, even in the light lunar gravity. The conveyor system was a bit balky—it had obviously never been tested in the lunar environment—and sprayed dust into the interior of the LM as Aldrin retrieved the samples.

Once this was done, Armstrong headed up the ladder. As he reached the hatch, Aldrin helped him maneuver inside—a complex task that Aldrin had just completed without assistance. "Just keep your head down close. Now start arching your back. That's good. Plenty

of room. Okay now, all right, arch your back a little. . . . Roll right just a little bit. Head down." Aldrin said.

Just over two and a half hours after they had opened the hatch, they had it closed again. "Okay, the hatch is closed and latched, and verified secured," Armstrong said.

FINAL MOMENTS ON THE MOON

As the LM repressurized with oxygen, the two astronauts noted, chuckling, that their pressure suits had not caught on fire. Long before departure from Earth, a physicist from Cornell University, Thomas Gold, had suggested that lunar dust on their suits might combust

A view from a Lunar Module's window, taken by Aldrin after the Moonwalk. Note the footprints throughout the shot. They will last for hundreds of millions of years.

A jubilant Neil Armstrong shortly after his Moonwalk. It was one of the few photos Aldrin took of him on the Moon, since Armstrong had the camera during most of the Moonwalk.

Buzz Aldrin looking tired but content after the Moonwalk.

when exposed to oxygen. He had also suggested that the Moon might be covered in such deep dust that the heavy LM could sink out of sight upon touchdown. Gold had been part of NASA's spaceflight planning since the late 1950s and had designed a close-up camera that was used on Apollo flights, so his opinions carried some weight at the agency—though few took these notions very seriously. In the end, neither hypothesis was correct. The astronauts did notice, however, that lunar dust—which ended up covering the interior of the LM in a fine silt—had a peculiar smell, one that Aldrin likened to spent gunpowder.

They had about thirteen hours to eat, rest, recuperate, and prepare to depart. Aldrin took a photo from the LM window of the work area below. "Houston. Tranquility Base. We're in the process of using up what film we have." Aldrin snapped off the remaining shots.

They rested briefly, then removed their helmets and gloves so that they could eat. The lunar surface menu included beef stew (dehydrated), bacon cubes, peaches,

coffee, date fruit cake, and more.

Soon they connected their suits to the LM's life-support system, put their helmets and gloves back on, depressurized the LM, reopened the hatch, and tossed out anything not needed to complete their mission. Out went the PLSS backpacks, out went the expensive Hasselblad cameras (some on Earth complained about this due to the cameras' cost, but compared to the backpacks and other discarded hardware, they were a pittance), out went anything not bolted down and essential to their ascent. The base of the LM soon looked like a pile of litter.

With the hatch closed, the pair repressurized the cabin once again and prepared to rest as best they could. Armstrong drifted into a light sleep fairly quickly, but Aldrin, closer to a noisy pump in the LM, had more trouble. The ascent from the Moon was just hours away, and they would need to be as alert as possible for it to be a success.

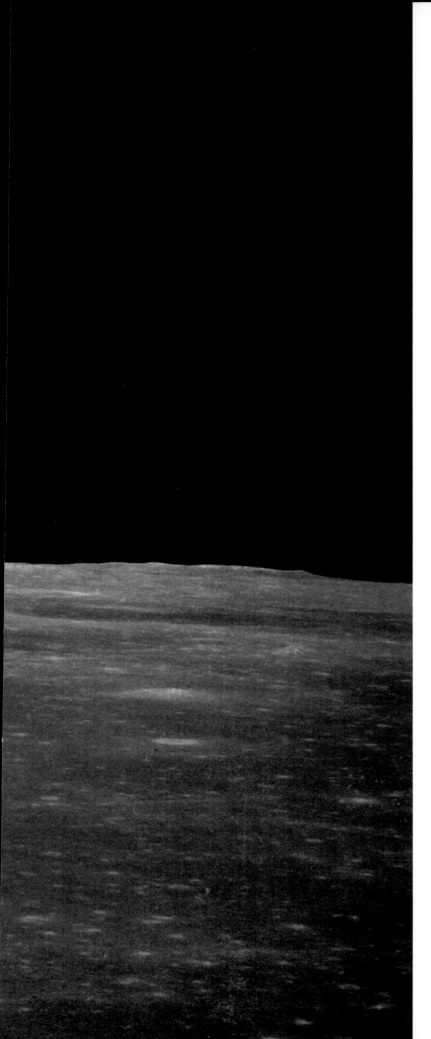

ONE CHANCE TO LIVE

Eight hours later, Armstrong and Aldrin were busy preparing to leave the Moon. Their rest had been fitful, and both were anxious to complete the mission. One task that was not on their pre-ascent checklist but was uncovered while working through it was the breaker switch that had been broken by Armstrong as he departed the LM hours earlier. This breaker needed to be closed to arm the ascent engine, and the discovery that it had been snapped off caused some understandable consternation at Mission Control. A bevy of technicians soon created a workaround that would have allowed the crew to arm the engine without the breaker, but in the end, the ever-practical Aldrin simply jammed the point of a felt-tip pen into the breaker to close it. The ascent engine was armed with a fix that had cost about thirty cents.

GOODBYE, MOON

With checklists complete and the LM ready to go (and the broken switch fixed), the duo was ready to leave the Moon. At 124 hours, 21 minutes into the mission, Aldrin counted down: "Nine, eight, seven, six, five, abort stage, engine arm, ascent, proceed." Armstrong keyed the computer and the little explosives below them opened the valves on the helium tanks, allowing the compressed gas to force the explosive chemicals into the ascent engine to do their job. The guillotine cut the connecting harness between the ascent stage and the descent stage, and the explosive bolts were severed. Both men breathed a sigh of relief as the ascent stage popped upward.

As *Eagle* streaked up toward its orbital rendezvous with Collins, Aldrin took a moment to look out the window. "We're off. Look at that stuff go all over the place," he said, referring to the bits of Mylar that scattered as they departed. "Look at that shadow. Beautiful."

Armstrong replied, "The *Eagle* has wings."

At just over 135 hours into the mission, Collins fired the engines on *Columbia* that broke the spacecraft free from lunar orbit and sent them onto a course for Earth.

Both men had noticed that the flag had been blown over by the ascent engine blast, but neither mentioned it on the radio. The LM lifted smoothly into the lunar skies, wallowing a bit as it did so—the fuels were held in tanks on either side of the cabin, and as they burned and the tanks emptied, *Eagle*'s balance went slightly off-center.

A bit over three hours later they were approaching *Columbia*. As the two spacecraft closed range, they were passing behind the Moon for the last time before they would dock. The Mission Control announcer said, "This is Apollo Control; 127 hours, 50 minutes ground elapsed time. Less than a minute now away from acquisition of the spacecraft *Columbia*. Hopefully flying within a few feet of it will be *Eagle*. Docking should take place about 10 minutes from now, according to the flight plan. However, this is a crew option matter. We're standing by for word that data is coming in from the two spacecraft."

As the two spacecraft neared each another, Collins kept Armstrong and Aldrin updated. He guided *Eagle*'s final maneuvers. "That's the way, keep going . . . go a little bit more . . . go ahead, go ahead . . . okay, stop. Okay, I got it now."

Armstrong told Collins that he was done maneuvering. "I'm not going to do a thing, Mike. I'm just letting her hold in attitude hold." *Eagle* would remain motionless while *Columbia* maneuvered toward it.

The two spacecraft docked. Collins unstrapped from his couch and drifted forward to open the CM's forward hatch and detach the docking probe—it had to be manually removed before Armstrong and Aldrin could reenter.

The two men passed the boxes of lunar samples, film magazines, and other paraphernalia to Collins, who squirreled them away in storage areas within *Columbia*. Two hours later they closed *Columbia*'s hatch for the last time and jettisoned *Eagle*. It drifted slowly away as they watched. After reporting technical data to Houston, Collins said, simply, "There she goes. It was a good one."

The three crewmen were reunited. To his relief, Collins's fears about the successful return of his crewmates turned out to be unfounded.

. . . AND GOODBYE, EAGLE

Back on Earth, Tom Kelly and his fellow Grummies felt a twinge. The first LM to land on the Moon had completed its task and passed with flying colors. The undocking represented the final act for the complex little ship, and Kelly could not have been prouder. He later reflected that the LM's contribution to the Apollo program "showed us what people could do in space under some very demanding conditions. And we learned an awful lot about the Moon, very interesting place, and gave us an idea about what it would be like to possibly explore other planets in the future as well. . . . I was delighted to work on it and really very happy that it went as well as it did."[64]

Five hours later, Collins set the computer to fire the CM's engine to break them free from lunar orbit and begin the long journey home. The crew was relaxed and joked about the upcoming engine firing.

As usual, Collins started the banter. "Yes. I see a horizon. It looks like we are going forward." They all laughed. "It is most important that we be going forward," Collins continued. He was referring to the fact that in project Gemini the retrofire maneuver that was used to slow them down and allow them to reenter Earth's atmosphere required the capsule to be pointed backward. Now, however, they wanted to make sure they were boosting forward to break free of lunar orbit.

"There's only one really bad mistake you can make there," Collins chuckled. Aldrin added, "Shades of Gemini retrofire, are you sure we're—no, let's see—the motors point this way and the gases escape that way, therefore imparting a thrust thataway."

Assured that they were heading back toward Earth, the trio settled in for the journey home.

OPPOSITE: Shortly before docking, Mike Collins snapped a photo that would soon be in newspapers and on magazine covers worldwide: the Earth rising over the Moon behind *Eagle*.

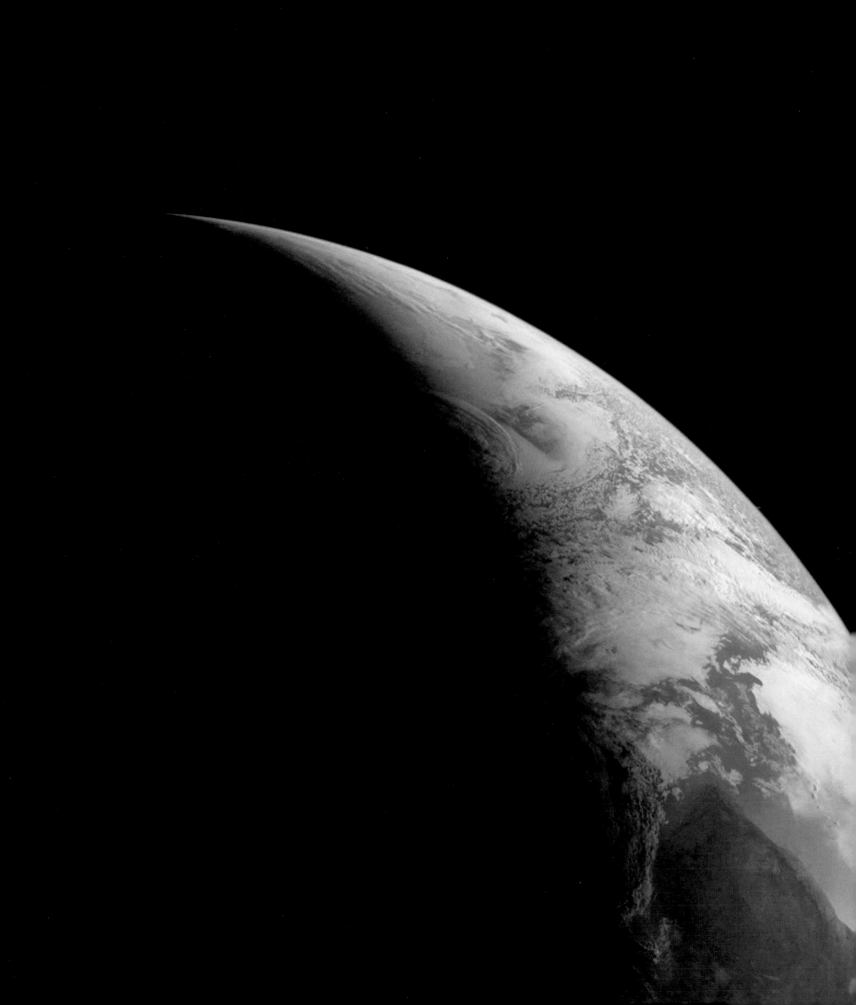

CHAPTER 12

COMING HOME

"ONLY FOUR NATIONS—COMMUNIST CHINA,
NORTH KOREA, NORTH VIETNAM, AND ALBANIA—HAVE NOT
YET INFORMED THEIR CITIZENS OF YOUR FLIGHT
AND LANDING ON THE MOON."

—Bruce McCandless, Apollo 11 CAPCOM

COLLINS WAS BACK IN THE DRIVER'S seat. There were midcourse correction burns to be completed to make sure they were flying the proper trajectory to enter Earth's atmosphere perfectly—the slightest misalignment could result in the spacecraft bouncing off the atmosphere, dooming them, or entering too steeply, resulting in potentially crippling g-forces during reentry.

At 148 hours into the mission, Bruce McCandless, the CAPCOM, relieved some of the ennui of the return flight with a news update. "From the hot wires of the Public Affairs Office, Apollo 11 still dominates the news around the world. Only four nations—Communist China, North Korea, North Vietnam, and Albania—have not yet informed their citizens of your flight and landing on the Moon." He continued, "The armed forces radio and TV network in Vietnam gave the mission full coverage." After updating a number of sports scores, McCandless added, "Mario Andretti won the 200-mile [322-km]

Trenton Auto Race, Sunday, and is now the leading race driver in the US Auto Club's point standings. And that's about the summary of the morning news this afternoon in Houston. Over."

Armstrong replied jokingly, "Look up the Dow Jones Industrials for us."

Columbia continued to sail the silent sea between the Moon and the Earth.

A few hours later, Mission Control began asking questions about the samples that had been collected from the Moon. "With respect to the documented sample container—on television it appeared to us as though the samples for that container were in fact being given—being selected in accordance with some thought or consideration being given to the

OPPOSITE: As they neared Earth, the astronauts were able to see specific features on their home planet. In this shot, the coast of Somalia is visible to the lower right.

Gene Kranz's White Team, the group that guided Apollo 11 to the surface of the Moon. Kranz is seen at center, with CAPCOM Charlie Duke to the right of him.

rocks themselves. And we were wondering if you could give any further details from memory about any of these samples, and the context of the material or the surface from which they were taken. Over."

Armstrong replied, "Yes. You remember I initially started on the . . . side of the LM that the TV camera was on, and I took a number of samples of rocks on the surface, and several that were just subsurface—15 to 20 feet [5 to 6 m] north of the LM. And then I recalled that that area had been probably swept pretty well by the exhaust of the descent engine, so I crossed over to the southern side of the LM and took a number of samples from the area around the elongate double crater that we commented on, and several beyond that, and tried to take as many different types of rock as I could see by eye, as I could in the short time we had available.

There were a number of other samples that I had seen earlier in our stroll around the LM that I had hoped to get back and pick up and put in the documented sample, but I didn't get those and I'll be able to comment on those in detail when we get in the debriefing session."

McCandless said, "Okay. Thank you, Neil. That about wraps up the questions we have on hand for now." There would be countless more from the geologists upon their return home.

About five hours before reentry, the CAPCOM on shift, astronaut Ron Evans, uplinked another news report—he called it the *Morning Bugle*. "President Nixon surprised your wives with a phone call from San Francisco just before he boarded a plane to fly out to meet you. All of them were very touched by your television broadcast. Jan and Pat watched from Mission Control here." He continued, "[Former astronaut] Wally Schirra has been elected to a five-year term on the board of trustees of the Detroit Institute of Technology. He will serve on the institute's development committee. Air Canada says it has accepted 2,300 reservations for flights to the Moon in the past five days. It might be noted that more than 100 have been made by men for their mothers-in-law. And finally it appears that rather than killing romantic songs about the Moon, you have inspired hundreds of song writers. Nashville, Tennessee, which probably houses

Collins mugs for the TV camera on the return voyage. He had begun growing a mustache during the mission.

the largest collection of recording companies and song publishers in the country, now reports it is being flooded by Moon songs. Some will make it. The song at the top of the best-sellers list this week is 'In the Year 2525.' *Morning Bugle* out."

The crew thanked Houston and continued to prepare for the critical reentry. They were approaching Earth at just under 10,000 miles (16,000 km) per hour.

REENTRY AND RECOVERY

Just prior to reentry, Collins turned the spacecraft to the proper orientation and released the Service Module from the Command Module. The 13-foot-high (4-m) *Columbia* would be the only part of the 363-foot (111-m) Saturn V rocket to return home, and reentry was just minutes away.

The capsule plunged into the Earth's atmosphere on July 24th. *Columbia*'s heat shield partially burned away as planned, taking much of the 5,000-degree-Fahrenheit (2,760°C) heat of reentry with it. This was only the third return from the Moon, and the speeds involved were much higher than with orbital flights, resulting in higher temperatures. Adding to the stressful wait in Mission Control, the plasma cloud generated around the Command Module prevented communication with the ground during this phase of the flight.

Collins recalls the experience as blinding: "The intensity of illumination has increased dramatically, flooding the cockpit with white light of startling purity. . . . We seem to be in the center of a gigantic electric lightbulb, a million watts' worth at least."[65]

All anyone outside the spacecraft could do was sit and wait, while anxious sailors aboard the recovery ships stationed in the Pacific Ocean scanned the skies for the three parachutes that should soon deploy to ease the descent of the CM. Apprehension filled the minutes as they ticked by.

Six minutes later the parachutes were sighted. Three minutes after that, Armstrong was back on the radio, responding to prompts from Mission Control—everything was okay. They hit the water right on time, and recovery divers were dispatched from a helicopter to fit floatation devices to the capsule. Apollo 11 had assumed the "Stable 2" position in the ocean, bobbing inverted in the swells, with the three astronauts hanging from the straps of their seats. Within minutes, floatation bags had righted the capsule. The astronauts took some antinausea pills anyway—not so much against seasickness now, but for when they would be buttoned up inside their protective biocontamination suits—vomiting inside those would be horrible.

Collins radioed the recovery forces, "This is Apollo 11. Tell everybody, take your sweet time. We're

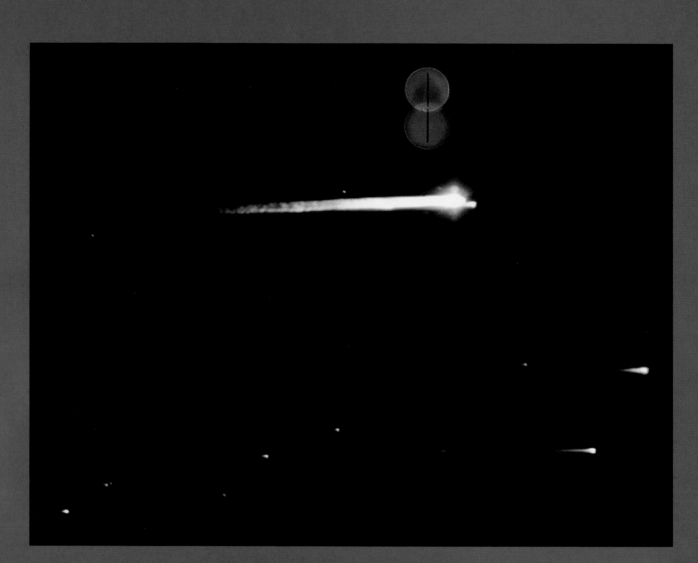

NASA-S-66-11003

ENTRY INTO EARTH ATMOSPHERE

ABOVE: The Apollo 8 capsule reentering the Earth's atmosphere. With temperatures up to 5,000 degrees Fahrenheit (2,760°C) greeting spacecraft returning from the Moon, a lot could go wrong. Fortunately, nothing did.

LEFT: A period illustration of an Apollo capsule reentering the Earth's atmosphere. If the heat shield had failed, the crew would have perished.

TOP: At home inside the modified Airstream quarantine unit. It was small, but still better than the Command Module with regard to comforts.

BOTTOM: The crew of Apollo 11 arrive at the aircraft carrier *Hornet* in their biological-isolation garments.

doing just fine in here." But the navy recovery team was working quickly to retrieve the astronauts.

Within minutes the hatch was opened and three bio-suits were tossed in—there had been concerns in the biomedical community that the crew might be carrying "Moon germs" that could be dangerous to life on Earth. The astronauts suited up and minutes later tumbled out of *Columbia* and into a waiting life raft.

Of course, any germs floating around inside the capsule could have drifted out into the atmosphere when the hatch was opened, dooming Earth to a microscopic alien invasion, but in the end it was ascertained that the trio carried only the germs they had taken with them when they left Earth.

LOCKDOWN

Soon, Armstrong, Aldrin, and Collins were retrieved from the bobbing capsule via a navy helicopter and transported to the aircraft carrier *Hornet*. There they disembarked wearing their drab-green bio-suits, and then were immediately walked to a modified Airstream travel trailer, their home for the next five days.

They would remain in quarantine for three weeks, and while the quarters were compact, they were palatial compared to what the trio had been used to for

the past eight days. Once they settled in, a doctor and cleaning technician joined them inside their temporary lockup. Then, with a bit of pomp and circumstance, President Nixon showed up outside the window of the Airstream to welcome them back to Earth—he had been flown out to the carrier to greet the astronauts. It would be many days until they were returned to Houston—still inside the modified trailer—and could see their families on the other side of the glass.

Soon the *Hornet* set sail for Hawaii, where the three astronauts and the two NASA specialists with them would be flown to Houston and transferred to the much larger quarantine facility at the Johnson Spaceflight Center.

JUBILATION IN HOUSTON

As the three astronauts settled into quarantine, people were celebrating at all the NASA facilities, but perhaps none with such utter exhaustion and elation as the men and women who had guided the mission from liftoff to splashdown from the Manned Spaceflight Center in Houston. Out came little American flags and cigars—Mission Control may have smelled like a three-day poker game, but the scene was colorful and buoyant.

Gene Kranz said of the aftermath: "I was more teary-eyed in the months after Apollo than at any time in my life. Every time I heard the National Anthem, or looked at the Moon, or thought of my team, I got misty."[66] This was not normal behavior for the ex-Marine. He had been transformed by the experience.

Sy Liebergot, the White Team EECOM, said, "We didn't consider them recovered until we saw the astronauts removed from the Command Module in the ocean by helicopter and step out on the deck of the aircraft carrier. Only then did we celebrate." He added, "We were young and we were fearless and, after all, nobody had ever told us young engineers that we couldn't successfully land humans on another planet. So we did it."[67]

TOP: *Columbia* arrives on the deck of the *Hornet*. Collins later said that the spacecraft had endured the rigors of the long journey better than he had expected.

BOTTOM: President Nixon welcomes the crew home shortly after their arrival on the *Hornet*.

TOP: Pass the cigars; senior Apollo management lights up in the post-splashdown celebration. At center, smiling, is Chris Kraft, director of flight operations at the Manned Spaceflight Center, and to right is Robert Gilruth, the director of the facility.

LEFT: Jubilation in Mission Control upon the conclusion of the mission of Apollo 11.

ABOVE: Breaking out the celebratory cigars at Mission Control. This was 1969, and not only were people allowed to smoke indoors, they had ashtrays at the ready on most control consoles.

AFTERMATH

"HISTORY IS A SEQUENCE OF RANDOM EVENTS AND UNPREDICTABLE CHOICES, WHICH IS WHY THE FUTURE IS SO DIFFICULT TO FORESEE . . . BUT YOU CAN TRY."

—Neil Armstrong, Apollo 11 astronaut

COMING HOME FROM A LONG WORK trip is never easy. Coming home from a working exploration of the Moon is indefinitely harder. And after everything else—being cooped up in 218 cubic feet (20 cu. m) of Command Module with two other guys who hadn't showered for over a week and a bunch of stinky rocks—the crew of Apollo 11 had to fill out customs forms. Yes, that return from Earth's nearest neighbor earned them the distinction of having been in foreign territory, and they had to deal with US customs. As a part of that, the trio had to list what they were bringing back to the United States, so they dutifully filled out an Agriculture, Customs, Immigration and Public Health General Declaration form. Departure: Moon. Arrival: Honolulu, Hawaii. Cargo: Moon rock and Moon dust samples.

After three weeks of confinement, the final leg of which was spent in the Lunar Receiving Laboratory in Houston, the astronauts were released to the embrace of family and friends. Before they could be released, however, an Interagency Committee on Back Contamination had to convene at the Centers for Disease Control and Prevention in Atlanta to lift the quarantine.

Upon reflection, the quarantine was not entirely wasted time. The astronauts welcomed the quiet to decompress from the mission before facing the world press. They were able to start writing their mission reports, and did so in relative calm.

During the early part of the quarantine, *Columbia* was attached to the isolation unit by a plastic tunnel. Collins quietly wormed down the tunnel at one point to make a visit to his home of eight days. Pulling out a felt-tip marker, he wrote on the control panel, "Spacecraft 107—alias Apollo 11—alias *Columbia*. The best ship to come down the line. God Bless her. Michael Collins, CMP." His sentiments were welcome, though perhaps not entirely accurate.

OPPOSITE: The August 13, 1969, parade down Broadway in New York City. Sitting high in the back seat, left to right, are Aldrin, Collins, and Armstrong.

ABOVE: The astronauts speak with their wives from their isolation quarters after landing in Hawaii.

LEFT: The customs form filled out by the astronauts upon their return to Earth. In the box questioning what diseases might be carried, they wisely wrote "to be determined."

Save for the explosion-stricken CM of Apollo 13, all the Command Modules performed flawlessly during their missions.

Toward the end of the crew's isolation, Aldrin was watching taped news coverage of the mission. It was strange to see it from an Earthly perspective—he had *been* there, but this felt, somehow, different. Aldrin later said that this was the first time he actually felt the emotional impact of that they had done. After one viewing session, he turned to Armstrong, who was silent in his own musings of the mission, and said, "Neil, we missed the whole thing." Armstrong smiled.

After twenty days the astronauts were released from their lockdown—the lab mice that had been sequestered with them had suffered no ill effects, and the blood tests were fine. The astronauts exited the quietude of quarantine and stepped into a maelstrom of activity that lasted for months.

On August 13, NASA sent the three astronauts on what can only be described as a whirlwind tour—in a single day, they participated in parades, meet and greets, and formal dinners in New York, Chicago, and Los Angeles. In New York confetti rained down from the sky, and in Los Angeles the crew and their families dined with President Nixon; his wife, Pat; Ronald Reagan, then governor of California; fifty members of congress; and representatives of eighty-three countries. The Century Plaza Hotel had never seen anything like it.

The astronauts then departed on a forty-five-day tour of twenty-five countries. In each capital they were given a hero's welcome and feted with great ceremony. It was a blur of activity that exhausted them and temporarily prevented each man from completing his final transition back to Earth.

HOME AT LAST

And then it was over. Not the adulation—that would last, in varying forms, for a lifetime. But the numbing string of speaking engagements, handshakes with foreign dignitaries, and kissing of babies slowed to a trickle. Each man was able to return to his home in Houston to ponder the future. Their paths ahead were forged by individual reactions and needs, and could not have been more different.

Neil Armstrong had decided long before that Apollo 11 would be his swan song at NASA. He'd spent enough time in training and in space, and enough time at the space agency. He felt that it was also his responsibility to step aside and allow other astronauts to climb the ladder—that's the kind of person he was.

Eager to escape the limelight, Armstrong became a professor of aeronautical engineering at the University

Neil Armstrong and John Glenn celebrate the fiftieth anniversary of Glenn's Mercury flight.

of Cincinnati in 1971. He took readily to the quiet anonymity of academic life and eschewed the hounding of the press. Armstrong said that he simply wanted to be treated as "Mr. Average Guy," wrote Al Kuettner, a representative of the university. Ron Huston, a fellow professor, said, "Neil viewed himself as just an ordinary person. . . . He fully understood that the Moon landing was the result of long, hard work of many people. Neil did not want to leave the impression that he did it all on his own."[68]

Of his work at the university, Armstrong once said, "I take substantial pride in the accomplishments of my profession. Science is about what is; engineering is about what can be."[69] He died of complications from cardiovascular surgery in 2012.

Mike Collins was not as shy as Armstrong, yet neither did he seek attention for himself. He too left NASA but eventually took a post with a very public-facing status: the director of the Smithsonian's National Air and Space Museum (NASM). Prior to that, he'd spent a year with the Department of State but found that the Smithsonian position suited him better. At NASM, he oversaw a massive expansion in the facility that was to open concurrent with the American bicentennial in

Mike Collins shares memories with Buzz Aldrin at Neil Armstrong's memorial service in August 2012.

Buzz Aldrin discusses the future of human spaceflight with Jeff Bezos at the National Space Society's International Space Development Conference® in 2018.

1976. In 1980 he left the Smithsonian to enter business. Now retired, Collins enjoys good food, fine wine, and watercolor painting. His artwork sells for thousands of dollars.

Buzz Aldrin's trajectory has been very different. Increasingly, he was aware that America's post-Apollo planning was on a tight budget, and this did not sit well with him. Not only had Richard Nixon cancelled the last three Apollo missions—18, 19, and 20—but he had scotched von Braun's plans for a human Mars mission. After the Apollo-Soyuz Test Project in 1975, the remaining Apollo hardware—built, flight-tested, and proven—was hauled off to museums such as Collins's Smithsonian. The Nixon administration demanded a cheaper, less adventurous approach to space exploration and sought this with the space shuttle program. In the end, it was an expensive and sometimes dangerous program, but it did result in the International Space Station.

Aldrin later became somewhat of a media darling, appearing on everything from *The Carol Burnett Show* in 1972 to *The Big Bang Theory* in 2012, with well over 150 media appearances in between. He has unceasingly and untiringly spoken strongly for a robust return to space beyond Earth orbit with an emphasis on Mars and international cooperation. Says Roger Launius, space-history curator at NASM, "He doesn't stop. He could if he wanted."[70] But he hasn't. With ongoing work in media, STEM education, and continuing outreach, Aldrin is a one-man advocacy organization for spaceflight. He sits on numerous boards, is a member of many professional organizations, and has served on the board of governors for the National Space Society for decades, working to advance this cause.

BACK TO THE MOON

"MEN MIGHT AS WELL PROJECT A VOYAGE TO
THE MOON AS ATTEMPT TO EMPLOY STEAM NAVIGATION
AGAINST THE STORMY NORTH ATLANTIC OCEAN."

—Dr. Dionysus Lardner, professor of natural history
and astronomy, University College, London, 1838

AH, THE NAYSAYERS. SO MANY LACK the vision to push beyond humanity's self-imposed barriers. But we have just started our relationship with the Moon—we've reached its orbit, descended to and explored its surface, and studied samples of it for decades. But that's just a beginning. Recent years have seen a renewed focus on orbiting scientific probes to gain further understanding of the Moon's makeup, and China has dispatched robotic landers and rovers to explore its surface anew. The near future holds promise for public-private partnerships in lunar development, both for its exploration and settlement and for moving beyond. The flights of Apollo took those first fledgling steps toward new worlds.

From 1969 through 1972, five more missions would visit the lunar surface with increasingly ambitious agendas, culminating with the magnificent flight of Apollo 17 which ended on December 19, 1972. The Apollo program taught us much. As a nation, it taught us that we can do what we set

10009,0

OPPOSITE: The Moon will soon become a hub of human activity, with NASA planning an orbiting station for the mid-2020s and Europe, China, and various private enterprises aiming for human-staffed surface activities later in the decade.

ABOVE: A large rock sample collected at the Sea of Tranquility.

TOP: What all the fuss was about: some of the Moon rocks returned by Apollo 11. By looks alone, they could be found in any backyard, but they are truly unique. Lunar samples may not be sold or owned by private individuals.

LEFT: The NASA *Apollo-11 Lunar Sample Information Catalog* entry on sample number 10009—a description only a scientist is likely to cherish.

our minds to. As a species, it taught us that there is more that unites us than divides us. As individuals, it taught us each something different. The lunar rocks themselves keep teaching us something new, year after year. They have been continuously studied at NASA and in universities and laboratories all over the world, ever since their delivery to Earth.

Besides coming to better understand the origins and history of the Moon from these lunar specimens, we have also learned of a possible future for humanity beyond Earth, for lunar rocks and soil are full of promise. While they appear to be gray and lifeless, they are

Both NASA and private industry have plans to mine the Moon and utilize its resources. Experts feel that the most likely path to success is via public-private partnership.

chock-full of usable resources that can expand human presence into and throughout the solar system.

The process of using lunar resources is broadly referred to as *in-situ resource utilization*, or ISRU. In brief, it is the use of materials found in space—whether in asteroids, in comets, on Mars, or, in this case, on the Moon—to further human endeavor. Anything that we don't have to carry to the Moon because it is found there saves energy and, in turn, money. This is not a new idea, but it was given new life when the rocks from the Apollo missions were analyzed. One of the most fascinating discoveries was the presence of water in Moon rocks, left over from its formation billions of years ago. This water can be purified for drinking and processed to make oxygen, a prime element of rocket fuel, and hydrogen can be processed as well. The oxygen is also

useful for breathing, of course. This means that instead of spending thousands of dollars per gallon to launch water, oxygen, and rocket fuel to take us beyond Earth orbit, these commodities can be mined, processed, and stored on the Moon and in Earth orbit to facilitate moving farther into the solar system.

Additionally, a number of metals are in the lunar surface. Iron, titanium, and aluminum are all present in abundance and can be extracted from lunar soil. This would facilitate the building of everything from lunar habitations to spacecraft—again, without the expense and complication of launching these heavy materials from Earth.

There is also helium-3 on the Moon, an element that is rare on Earth, which could ultimately prove useful for nuclear fusion reactors in the future.

The Moon Village, a lunar outpost proposed by the European Space Agency and to be built in association with international partners.

NASA's Lunar Orbital Platform-Gateway, sometimes shortened to LOP-G or just Gateway, is scheduled for construction in the early 2020s.

CAPITALISM ON THE MOON

There are a number of commercial companies working on lunar-mining technologies and the means for transporting the end result, including Blue Origin, the aerospace company founded by Jeff Bezos. With massive amounts of private investments pouring in for these efforts, it is just a matter of time before "Made on the Moon" becomes a moniker affixed to all kinds of machinery and merchandise found in space.

That is just one example of the plans that could enable a return to the Moon, and most of them have only become practical in the last decade. Companies like SpaceX have paved the way for private investment in space ventures, suggesting to bold investors that there is money to be made not only in space, but

specifically on the Moon. Such investment is currently at an all-time high.

But these are merely reasons for going back. Actually committing to the endeavor has been far harder, especially where humans are concerned. Despite a very successful run of lunar exploration in the 1960s and 1970s, NASA has continually been frustrated regarding its efforts to return humans to the Moon. Not that the space agency hasn't tried—at least two major human-crewed lunar programs have been announced since Apollo, each touted by a presidential administration, but in both cases, the funds were never delivered by Congress to follow through. The most recent was NASA's Constellation program, pushed forward by George W. Bush in 2004. Constellation stumbled along

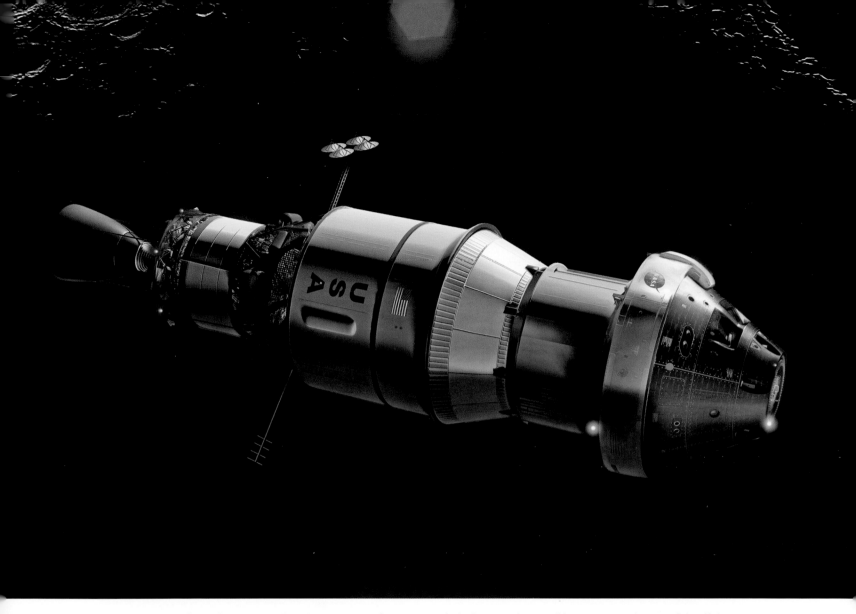

At some point in the early 2020s, NASA's Orion spacecraft, a version of which is seen here, will be sent into a looping flyby of the Moon atop the Space Launch System rocket. It's a less daring version of the flight of Apollo 8, but a necessary test of the hardware destined to return Americans to the Moon.

for six years, continually underfunded, until being terminated by President Obama in 2010 after a high-level study showed the plan to be infeasible with the level of investment specified.

The latest iteration of an American return to the Moon has come from the Trump administration, which made a lunar return a priority in 2017. Current plans involve an orbiting lunar station, called the Lunar Orbital Platform-Gateway, which is supposed to begin construction in the early 2020s. Whether or not this will ultimately succeed has much more to do with politics than capability—America went to the Moon with 1960s technology, and returning there will be much easier with a half century of technological advancements in hand. It is simply a matter of political will and national priorities.

Other nations also have plans for the Moon, China paramount among them. The rising space power has already sent landers and rovers there, and it plans to land a few of its citizens on its surface in the 2030s. The European Space Agency, a consortium of European nations, has plans for a "Moon Village," built in association with the US, Russia, and possibly China. And the private companies mentioned earlier are planning not just lunar mining, but crewed lunar bases as well—it's all a matter of investment and the creation of a marketplace that will enable them to make a profit in their pursuits. NASA and the US government will likely be a part of any such undertaking, whether via direct investment, creative public-private partnerships, or the creation of a commodities market in space.

"We came in peace for all mankind. July 20, 1969."

FINAL WORDS

In an interview in 2011, Armstrong advocated for a return to the Moon: "I favor returning to the Moon. We made six landings there and explored areas as small as a city lot and perhaps as large as a small town. That leaves us some 14 million square miles [36 million sq. km] that we have not explored." He added that this would allow engineers to practice "a lot of the things that you need to do when you are going further out in the solar system."[71]

At the end of a 2005 interview about Apollo 11 for a television show, I asked Gene Kranz if he had a final comment he would like to make. In most such interviews, the person in front of the camera says, "No, that's about it, I guess." But not Kranz. He paused for a moment, then looked up at me with his steely missile-man eyes and said with great conviction this simple but bold statement: "What America will dare, America can do."

I'm sure that Armstrong, Aldrin, and Collins would strongly suggest that we dare again and honor the great and everlasting legacy of Apollo 11.

ACKNOWLEDGMENTS

Thanks are due to the many, many people who assisted in the creation of this book.

First, thanks to the wonderful crew at Sterling Publishing Company, Inc., from top to bottom. You are as good as it gets. Special thanks to Meredith Hale, Katherine Furman, Barbara Berger, Ashley Prine, and Christopher Bain.

To John Willig of Literary Services Inc., my agent of nearly a decade, my enduring gratitude. He is simply tops in the business and a true friend as well. I owe you about fifteen steak dinners now.

Artist James Vaughan put in countless hours creating the wonderful, fresh illustrations for this book, and for that he has my deep respect and appreciation. Chronicling the space race is a labor of love for me, and I am fortunate to have found a fellow soul who feels the same way and expresses it in such artistic terms. If you have a moment and want to browse some truly moving aerospace artwork of a kind you've not likely seen before, head over to www.jamesvaughanphoto.com and prepare to be dazzled.

A tip of the hat to the multitalented Francis French, a gifted space author, presenter, and all-around raconteur who is also the director of education at the San Diego Air and Space Museum. He writes books with Apollo astronauts in his scarce spare moments, yet still occasionally finds time to do technical reads and edits of his colleagues' manuscripts. A humble and deeply heartfelt thanks to you, Francis, for lending me your brilliant mind and vast knowledge of the subject.

Nick Howes, a connoisseur of all things meteoritic and an avid space historian, provided the impossible-to-get image of a 1202 computer alarm on the Apollo Guidance Computer display from his very own reproduction unit, bought at great expense and with programs hand-loaded and executed. Thanks, Nick.

A special thanks to Andy Aldrin and Rick Armstrong, two people who are asked endless questions about their fathers and answer them with unending patience and goodwill. Both of these smart, talented men have impressive life paths of their own, and I appreciate them taking the time to step back fifty years and relive some of the highlights of their experiences growing up.

My humblest thanks is due Buzz Aldrin for generously providing a foreword for this book. Your ongoing efforts to promote human spaceflight and inspire new generations of space-savvy youth are truly impressive, and, to this space geek, your ability to navigate Gemini 12 by hand was simply amazing.

Finally, to the half-million Americans who sent human beings to the Moon nine times, my deepest respect. Your creations were decades ahead of their time, and your perseverance was, and remains, an inspiration to an entire generation and beyond. Through books like this and others, it is my fondest hope that the many sacrifices you made to achieve President Kennedy's goal will not be forgotten.

NOTES

1. Interview with the author, June 2005.

2. Ibid.

3. Transcript of Apollo 11 landing with annotations. *Apollo Lunar Surface Journal*, https://www.hq.nasa.gov/alsj. Accessed September 2017.

4. Excerpt from a letter from Wernher von Braun to the vice president of the United States, dated April, 29 1961. NASA Historical Reference Collection, NASA Headquarters, Washington, DC.

5. "Recommendations for Our National Space Program: Changes, Policies, Goals." A report from James E. Webb and Robert McNamara to Vice President Lyndon B. Johnson, May 8, 1961.

6. John F. Kennedy. "Special Message to the Congress on Urgent National Needs." May 25, 1961. The American Presidency Project archives, University of California at Santa Barbara.

7. Kennedy, "Special Message."

8. James Hansen, *First Man: The Life of Neil A. Armstrong* (New York: Simon & Schuster, 2005), 221.

9. Buzz Aldrin and Malcom McConnell, *Men from Earth* (New York: Bantam Books, 1989), 69.

10. Aldrin and McConnell, *Men from Earth*, 103.

11. Michael Collins, *Carrying the Fire: An Astronaut's Journeys* (New York: Farrar, Straus and Giroux, 1974).

12. Collins, *Carrying the Fire*.

13. Gene Kranz. *Failure Is Not an Option: Mission Control from Mercury to Apollo 13 and Beyond* (New York: Simon & Schuster, 2000).

14. Hansen, *First Man*, 246.

15. Ibid.

16. Hansen, *First Man*, 265.

17. *NOVA*, season 41, episode 23, "First Man on the Moon," directed by Duncan Copp and Christopher Riley, aired November 29, 2014, on PBS.

18. Aldrin and McConnell, *Men from Earth*, 154.

19. Aldrin and McConnell, *Men from Earth*, 157.

20. Aldrin and McConnell, *Men from Earth*, 15.

21. Transcribed from CBS coverage of the launch of Apollo 4.

22. Courtney Brooks, James M. Grimwood, and Loyd S. Swenson Jr., *Chariots for Apollo: A History of Manned Lunar Spacecraft*, the NASA History Series (Washington, DC: NASA Special Publication-4205, 1979), "NASA-Grumman Negotiations."

23. Thomas J. Kelly, *Moon Lander: How We Developed the Apollo Lunar Module* (Washington, DC: Smithsonian Books, 2001).

24. Kelly, *Moon Lander*.

25. Kelly, *Moon Lander*.

26. NASA transcript of preflight briefing, July 5, 1969.

27. Collins, *Carrying the Fire*.

28. Collins, *Carrying the Fire*.

29. Collins, *Carrying the Fire*.

30. Aldrin and McConnell, *Men from Earth*.

31. Collins, *Carrying the Fire*.

32. Private correspondence published in the Apollo 15 flight journal, https://history.nasa.gov/afj/ap15fj/03tde.html. Accessed June 10, 2018.

33. Collins, *Carrying the Fire*.

34. In-flight transcript (here and throughout chapter), NASA History Portal, https://www.jsc.nasa.gov/history/mission_trans/apollo11.htm. Accessed June 1, 2018.

35. Post-flight debriefing-session audio transcript, 1969.

36. Aldrin and McConnell, *Men from Earth*.

37. Jennifer Bogo, "Landing on the Moon," *Popular Mechanics*, May 2009.

38. Bogo, "Landing on the Moon."

39. Bogo, "Landing on the Moon."

40. Bill Safire, "In Event of Moon Disaster." Dated July 18,1969, https://www.archives.gov/files/presidential-libraries/events/centennials/nixon/images/exhibit/rn100-6-1-2.pdf. Accessed June 6, 2018.

41. Interview with the author, 2005.

42. Ibid. Since this speech was not recorded, this is the best approximation he and his controllers recall; it is written differently in various sources, but the essence is the same.

43. Ibid.

44. Interview with the author, 2005.

45. Eric M. Jones and Ken Glover, eds., *Apollo Lunar Surface Journal*. NASA publication.

46. Kranz, *Failure Is Not an Option*.

47. Bogo, "Landing on the Moon."

48. Bogo, "Landing on the Moon."

49. Bogo, "Landing on the Moon."

50. Excerpt from a letter by Margaret H. Hamilton, director of Apollo Flight Computer Programming at MIT's Draper Laboratory, Cambridge, Massachusetts, March 1971.

51. Rick Houston and Milt Heflin, *Go, Flight!* (Lincoln, NE: University of Nebraska Press, 2015).

52. Bogo, "Landing on the Moon."

53. Hansen, *First Man*.

54. Interview with the author, 2005.

55. Post-flight debriefing-session audio transcript, 1969.

56. Andrew Chaikin, *A Man on the Moon: The Voyages of the Apollo Astronauts* (New York: Penguin Books, 1994).

57. In-flight transcript (here and rest of chapter). NASA History Portal, https://www.jsc.nasa.gov/history/mission_trans/apollo11.htm. Accessed June 1, 2018.

58. Hansen, *First Man*.

59. In-flight transcript. NASA History Portal, https://www.jsc.nasa.gov/history/mission_trans/apollo11.htm. Accessed June 1, 2018.

60. Interview with the author, 2005.

61. Jones and Glover, *Apollo Lunar Surface Journal*.

62. In-flight transcript. NASA History Portal, https://www.jsc.nasa.gov/history/mission_trans/apollo11.htm. Accessed June 1, 2018.

63. Post-flight debriefing-session audio transcript, 1969.

64. NASA oral history with Tom Kelly, www.jsc.nasa.gov/history/oral_histories/KellyTJ/KellyTJ_9-19-00.htm

65. Collins, *Carrying the Fire*.

66. Kranz, *Failure is Not an Option*.

67. Jennifer Bogo et al., "Apollo 11: No Margin for Error," *Popular Mechanics*, June 2009.

68. Deborah Rieselman, "Little-Known Insights Tell How One Small Step Led to a Reluctant Hero." *UC Magazine*, University of Cincinnati, undated, https://magazine.uc.edu/issues/0413/Armstrong.html. Accessed August 9, 2018.

69. Neil Armstrong, excerpt from a speech given at a National Press Club event, February 2000.

70. Jeremy Hsu, "How Astronauts Can Become Media Stars," Space.com, March 19, 2010, https://www.space.com/8067-astronauts-media-stars.html. Accessed August 9, 2018.

71. "Armstrong Urges Return to the Moon, Then Mars," Seeker (website), August 25, 2011, https://www.seeker.com/armstrong-urges-return-to-the-moon-then-mars-1765387080.html. Accessed August 9, 2018.

IMAGE CREDITS

INDEX

Page numbers in *italics* indicate photographs or diagrams.

Abort procedure, *95–97*, 115, 124, 127. *See also* In-flight emergencies

Agena rocket
  adaptation to Gemini program, 41–43
  Agena A (version), 42
  Agena D (version), 43, 45, *56*
  use on Gemini 6, 41
  use on Gemini 8, 43–47
  use on Gemini 10, 51–53
  use on Gemini 12, 54–57

Agnew, Spiro, *96*

Aldrin, Andrew ("Andy"), 58–60

Aldrin, Edwin E. ("Buzz")
  astronaut selection, 18–19
  background and experience, 22–23
  children/family, 58–60
  Gemini 12 crewmember, 41
  launch and flight to Moon, 97, 105
  LM descent and Moon landing, 2–4, 6–7, 108, 112–119, 122–127
  LM recovery/redocking, 156–161
  Moonwalk/surface activities, *102–103*, *120–121*, 128–156
  observations from backside of Moon, 108
  prelaunch activities, 92–95
  reentry, retrieval, and quarantine, 167–168, 172
  retirement and life after Apollo, 174–175
  return to the Moon, advocacy of, 184

Anders, Bill, 46, 93

Apollo 4, *62*, 76–77

Apollo 5, *111*

Apollo 8, 3–4, *46*, 76, 89, 91, 109, 130, *166*

Apollo 9, 91, *100*, *104*

Apollo 10, 3–4, 91, *114*, 124

Apollo 12, 76, *133*

Apollo 13, 46, 89, 172

Apollo 16, 7, *86*, *131*

Apollo 17, 63

Apollo 18–20, cancellation, 175

Apollo Guidance Computer (AGC), 7, 118–119, 122. *See also* Computers/computer capabilities

Apollo Lunar Surface Close-up Camera (ALSCC), *102*, *121*

Apollo Portable Life Support System (PLSS), 131, 157

Apollo-Soyuz Test Project, 146, 175

Armstrong, Eric Alan ("Rick"), 60–61

Armstrong, Jan, 165

Armstrong, Neil A.
  astronaut selection, 18–19
  background and experience, 20–21
  children/family, 58, 60–61
  Gemini 8 crewmember, 41
  launch and flight to Moon, 97
  LM descent and Moon landing, 108, 122–127
  LM recovery/redocking, 156–161
  LM simulator training, *xiv*, *113*
  Moonwalk/surface activities, *102–103*, 128–156
  prelaunch activities, 92–95
  reentry, retrieval, and quarantine, 167–168, 172
  return to the Moon, advocacy of, 184
  teaching career and death, 173–174

Atlas rocket (US), 13, *34*, *42*, 70

Bales, Steve, 108–109, 118, 122, 124

Ballistic missiles, 10, 28, 66

Bell Aerosystems, 80

Bezos, Jeffrey P. ("Jeff"), 180

The Big Bang Theory (TV program), 175

Blue Origin, 179–180

Borman, Frank, 109

Bush, George W., 182

Buzz Aldrin Space Institute, 59

CAPCOM. *See also* Mission Control
  Apollo 11 flight, 2–3, 6–7
  Apollo 11 launch, 99
  Apollo 11 LM landing, 109, 115, 117–119, 122–123
  Apollo 11 LM surface communication, 146, 152–153, 155
  Apollo 11 return flight, 163–165
  Gemini 8, 44–47

Cape Canaveral, 43, 54, 77, 95–97, 99

Carlton, Bob, 109

The Carol Burnett Show (TV program), 175

Celestial/manual navigation, 51, 55, 57, 107

Centaur (rocket stage), 70

Cernan, Eugene A. ("Gene"), 63, 91

China, 163, 177, 183

Cold War, US–Soviet space race, 11–15, 41, *99*

Collins, Michael
  astronaut selection, 18–19
  background and experience, 23
  children/family, 58
  crewmember on Gemini 10, 39, 41, 51–53

EVA activities, 23, 51, 53
   launch and flight to Moon, 97
   LM descent/undocking, 112–114
   LM recovery/jettisoning, 159–161
   piloting the CM, 2–3, 79, 91, 100–101, 104–105,
      107–108, 155–156
   prelaunch activities, 92–95
   reentry, retrieval, and quarantine, 165–168, 171–172
   retirement and life after Apollo, 174–175
   return to the Moon, advocacy of, xii, 184
Collins, Pat, 165
Columbia (Jules Verne spacecraft), 26, 92
Columbia (NASA spacecraft). *See* Command Module
Columbus, Christopher, 92
Command Module (*Columbia*, CM)
   communication with LM, 2–3, 150
   development and naming, 34–37, 64, 83, 92
   electronics and computer controls, 76
   LM descent/undocking, 112–115, *150*
   LM docking probe, *101*, 161
   LM recovery, *91*, 99–101, 161
   lunar orbit, insertion into, 107–108
   lunar orbit, release from, 161
   navigation and trajectory control, 107, 112
   Passive Thermal Control (PTC), 105
   preflight check and countdown, 95–97
   reentry and splashdown, 165–168
   separation from S-IVB stage, 99–101, 104
   views, in-flight, *4, 104–105*
   views, interior/control panels, *96, 105*
   weight and size, 64
Computers/computer capabilities. *See also* Apollo Guidance
   Computer
   development, 22, 80–81
   Gemini 8, 43–47
   Gemini 12, 54–55
   guidance and navigation systems, 75–76, 161
   linkage to Mission Control, 6–7
   LM-CM DSKY interface, 108, 115, 118–119
   manual override of LM landing, 122–124
   rocket engine design, 68, 75
Constellation program, 182
Cronkite, Walter, xi, 77

Dietrich, Chuck, 109
Direct Ascent, lunar landing, 29–32, 34–37, 63
Docking. *See* Rendezvous and docking
Dolan, Thomas, 35
Douglas Aircraft Company, 21, 74
Duke, Charles M., Jr. ("Charlie"), 2, 6, 7, 115, 117,
   122–127, *130*, 164

Eagle. *See* Lunar Module
Early Apollo Scientific Experiment Package (EASEP), *148,*
   149, 152–153
Earth Orbit Rendezvous (EOR), lunar landing, 31–32,
   35–37
Earth, views from space, *158–160, 162*
Edwards Air Force Base, 19–21, 23, 69
Eisenhower, Dwight D., *x*
Equigravisphere, 107
European Space Agency (ESA), 183
Explorer 1 rocket (US), *13*
Extra vehicular activity (EVA), 23, 38–41, 51–57, 82,
   153–155
   EVA suits, *60*, 94–95, 131, 140, 156–157

Faget, Maxime, 28, *29*, 37
From the Earth to the Moon (Verne), 26–27, 92

Gagarin, Yuri, 14
Garman, Jack, 109, 118
Gavin, Joe, 109
Gemini program
   capabilities and accomplishments, 39–41
   Gemini 1, 41
   Gemini 2, 41
   Gemini 3, 41
   Gemini 4, *38*, 41
   Gemini 5, 41
   Gemini 6/6A, *40*, 41
   Gemini 7, *40*, 41
   Gemini 8, 41, 43, 45–46, *50*, 51, 61
   Gemini 9, 51
   Gemini 10, 23, 39, 41, 51–53
   Gemini 11, 41, 53
   Gemini 12, 22, 41, 54–57, 59
   origins/development, 21
Gilruth, Robert, *169*
Glenn, John, 23, *173*
Goddard, Robert, 28
Grumman Aircraft Engineering, 80–89, 109, 127
Grumman F9F Panther (jet fighter aircraft), *19*, 83

Haise, Fred, *95*
Haldeman, H. R., *110*
Ham (chimpanzee), 13–14
Hamilton, Margaret, 119, *122*
Hamilton Standard (company), 80
Haney, Paul, 99, *132*
Hodge, John, 47, *50*
Holloman Aerospace Medical Center, 13
Houbolt, John, 27, 34–37
Houston, TX. *See* CAPCOM; Mission Control

Huntsville, AL, 21, 34, 68

Huston, Ron, 174

In-flight emergencies. *See also* Abort procedure
    Apollo 12 electrical problem, 76
    Apollo 13 explosion, 89, 172
    Command 400 (computer code), 45
    Gemini 8 loss of control and reentry, 45–47, *50*
    Gemini 12 loss of radar, 55
    LM 1202 Program Alarm, 117–119, 122
    LM EVA test, *82*
    LM surface liftoff, 127, 129, 140

In-situ resource utilization (ISRU), 178–179

Interagency Committee on Back Contamination, 171

Intercontinental ballistic missiles (ICBM), 10, 28, 66

International Space Development Conference, *175*

International Space Station (ISS), 175

Isabella (queen of Spain), 92

Johnson, Lyndon B., 15, *96*

Johnson Space Center (formerly Manned Spaceflight
    Center), 34, *58*, 108, 168

Kelly, Thomas J. ("Tom"), 82–84, 87, 89, 109, 127, 161

Kennedy, John F., xii, 9, 14–17, 19, 22, 29, 32, 34, 37, 66,
    89, 146

Kennedy Space Center, 4, 27
    astronaut quarantine, 93
    construction, 21
    Firing Room, 95–96, 99
    Vehicle Assembly Building, *33*

King, Jack, xi, *97*

Kondratyuk, Yuri, 35

Korean War, 19–20, 22, 83

Kraft, Chris, 111, *169*

Kranz, Eugene F. ("Gene"), 4, 32, 108–109, 111, 115,
    117–118, 122–124, 127, 130, *164*, 168, 184

Kuettner, Al, 174

Laika (Soviet dog), 13

Langley Research Center, 34

Lardner, Dionysus, 177

Laser Ranging Retro Reflector (LRRR), *149*, 152, 155

Launius, Roger, 175

Lewis Flight Propulsion Laboratory, 20

Liebergot, Seymour A. ("Sy"), 108, 168

Lockheed Aircraft, 21, 82

Los Angeles Air Force Base, 22

Lovell, James, 45–46, 54–57

"Lunar bunny hop," 60, 143

Lunar Equipment Conveyor (LEC), 138, 155–156

Lunar Module (LM, "lem")
    about the development, 64, 79–89

approval for descent, 2–7, 112–115
    CM docking probe, 100–101, 161
    dedication plaque, 92–93, *142–144*
    Familiarization Manual, 5
    landing/exiting/touching the surface, 6–7, 122–127,
        131–135
    Landing Point Designator, 6, *115*
    on the Moon's surface, *128–159*
    preparing for departure, 155–161
    redocking procedure, 38
    rendezvous and docking maneuver, 42, *160–161*
    retrieval and jettisoning, 161
    Transposition and Docking Maneuver (TDM), 99–101,
        104
    views, control panels/interior, *xiv*, *4*, *48–49*, *113*, *132*
    views, exterior/in-flight, *2*, *78*, *91*, *112–113*, *150*, *160*
    views, mounted in S-IVB stage, *100*, *104*
    weight and size, 49, 64, 79 "White Team," 108–109,
        164, 168

Lunar Orbiting Platform-Gateway (LOP-G), *182*, 183

Lunar Orbit Rendezvous (LOR), lunar landing, 34–36

Lunar photographs
    astronauts/LM on the surface, *128*, *133–134*, *136–140*,
        142–149, *151*, *153–154*
    landing site, *116*
    Moon rocks/sample collection, *177–178*
    overhead back side, *106*
    overhead near side, *28*
    Sea of Tranquility (Mare Tranquillitatis), *1*, *123*
    surface still shots, *24–25*, *60*, *102–103*, *120–121*, *130*, *184*
    US flag, *156*
    view of Earth from space, *158–160*, *162*

Lunar Receiving Laboratory, *18*, 171

Lunar Reconnaissance Orbiter, 1

Lunar Roving Vehicle, 143

Lunar Sample Information Catalog, 178

Lunar samples
    Armstrong's sampling method, 163–164
    bulk sample, 146–148
    contingency sample, 138–140
    core samples, 153–155
    documented sample, 153
    in-situ resource utilization, 178–179
    loading into LM, 155–156
    transfer to CM, 161

Man in Space Soonest program (MISS), 20

Manned Lunar Landing and Return (MALLAR), 35–37

Manned Spaceflight Center (renamed Johnson Space
    Center), 34, *58*, 108, 168

Marshall Space Flight Center, 21, 34, *65*, 68

Mars, mission planning, 175, 178

Massachusetts Institute of Technology (MIT), 22, 80, 119, 122
Maynard, Owen, 28
McCandless, Bruce, 99, 163–164
McDonnell Aircraft Corporation, 21
McDonnell-Douglas Aerospace, 74
McNamara, Robert, 15
Mercury program. *See* Project Mercury
Mission Control, 104, 108, 111, *114*, 125–127, 130–132, 135, 138, 161, 168–169. *See also* CAPCOM
Mississippi Test Facility (renamed Stennis Space Center), *71*
Moon, advocacy for man's return, 179–184
Moon rocks. *See* Lunar samples
Moon Village, 181, 183
Morning Bugle (CAPCOM news report), 164–165
Mueller, George, 76, 95, *99*

Nance, Bob, 124–125
National Aeronautics and Space Administration (NASA)
    application requirements, 22
    astronaut selection, "Mercury Seven", 14, 19, 23, 39
    astronaut selection, "New Nine", 19, 21, 39
    budget and workforce, 33–34, 65–66
    creation, 13
    Kennedy lunar landing mission, 14–17
    Shepard suborbital flight, 14
    symbolism of the lunar landing, 92–93
    Vehicle Assembly Building, *33*
National Air and Space Museum (NASM), 174–175
National Space Society, *175*
Nellis Air Force Base, 22
Nixon, Pat, 173
Nixon, Richard M., 109, 111, 146, 164, 168, 173, 175
North American Aviation, 21, 70, 72–73
Nova (rocket), 31–32, 34, 36, 63, 66

Obama, Barack, 182
Oberth, Hermann, 28, 35
O'Hair, Madalyn Murray, 130
Orion (spacecraft), *183*

Paine, Tom, 97
Passive Thermal Control (PTC), 105
Petrone, Rocco, *95*
Phillips, Sam, *95*, *99*
Photographs. *See* Lunar photographs
Program Alarm (1201/1202 error code). *See* Apollo Guidance Computer
Project Dyna-Soar, 20
Project Mercury
    building the capsules, 29
    development, 20–21, 29, 39–40, 63

fiftieth anniversary of Glenn's flight, *173*
flight of Ham (chimpanzee), 13–14
selection of the "Mercury Seven," 14, 19, 23, 39
suborbital flight by Shepard, 14–15, 23, 32
Project Vanguard, 11–13

Quarantine, astronaut
    preflight, 93
    splashdown, 167–168, 171–173

R-7 (Soviet missile), 11, 13
Raytheon Company, 80
Reagan, Ronald, 173
Redstone rocket (US), 11, 13–14, 32, 65
Religion and spaceflight, 130–131
Rendezvous and docking
    Agena rocket, 40–41
    debate over direct ascent vs. EOR, 31–34
    development of LOR concept, 34–37, 63
    Gemini 6/7 missions, 41
    Gemini 8 mission, 41–47
    Gemini 10 mission, 51–53
    Gemini 12 mission, 54–57
    LM retrieval, 100–101, *104*, *150*, 161
Rice University, 9, *17*, 34
Rocketdyne (company), 66, 68, 74–75, 99
Rockwell, Norman, 140
Russia, 183. *See also* Soviet Union

Safire, William, 109–111
Saturn rocket
    "all-up" testing, 76
    guidance and navigation, 75–76
    Saturn I, 28, *30*, 31–32, 34, 65–66, *74*
    Saturn IB, 43, 77
    Saturn V, *31*, *62*, 63–68, 70–77, 81, 86, 88, *91*, 96–99, 165
    Saturn C-1, *30–31*
    Saturn C-3, 37
    Saturn C-4, 74
    weight and size, 63–64, *65*
Saturn rocket engines
    combustion instability, 68
    cryogenic fuels, 70–72
    F-1, 64–71, 75, 77, 99
    H-1, 66
    J-2, 68, 70, 74–75, 99
    RL-10, 75
    trans-lunar injection burn (TLI), 75, 99
Saturn rocket stages
    engine thrust, 65–66
    flight profile, 64

S-I, 70
S-IC, 66
S-II, 64, 66, 70, *71*, 72, 75, 99
S-IV, 64, 71
S-IVB, 66, 74, 76, 99–101, 104
S-band antenna, *3*, *6*
Scheer, Julian, 92
Schirra, Walter M., Jr. ("Wally"), 165
Scott, David R., 41, 43–47, 50, 61, 101
Seamans, Robert, 35–37
Sea of Tranquility (Mare Tranquillitatis), *1*, *123*, 146, 157, *177*. *See also* Tranquility Base
Shepard, Alan, 14–15, 23, 32
Silver, Lee, 147–148
Simulation supervisors ("Sim Sups"), *4*, 118
Simulator training
    computer errors, 118
    docking procedure, 100–101
    in-flight emergencies, 47
    LM landing, *xiv*, *4*, *7*, *59*, 109, *113*
    zero-g, *54*, *57*
Skylab (space station), 146
Slayton, Donald K. ("Deke"), 20, 22–23, 93, 122–123, *132*
Smithsonian Institution, 174
Solar Wind Collector, 144
Soviet Union
    destruction and rebuilding from WWII, 9–10
    space program accomplishments, 11–13, 41
    US space race, xi, 13–15, 20
Spacewalks. *See* Extra vehicular activity
SpaceX, 182
Special Message to the Congress on Urgent National Needs (Kennedy address), 16–17
Sputnik/Sputnik 2, 11, 13
Stafford, Thomas P. ("Tom"), 51, 91
Stennis Space Center (formerly Mississippi Test Facility), *71*

Television broadcast of Apollo 11, xi–xii, *134*, *144*, *146*, 164–165
Thompson Ramo Woolridge (TRW Inc.), 80, 127
Titan missile, 43

Tracking stations, 45–46, 51, 135, *139*
Tranquility Base
    "*Eagle* has landed" 125–127
    "*Eagle* has wings" 159–161
Transposition and Docking Maneuver (TDM), 99–101
Trump, Donald J., 183
Tsiolkovsky, Konstantin, 28
1201/1202 Error code (Program Alarm). *See* Apollo Guidance Computer

Underwater training, 54–55, 57, 59
University of Cincinnati, 173–174
US Air Force, 13, 20, 22, 23, 43, 76, 82, 83
US Centers for Disease Control (CDC), 171
US Customs Service, 171–172
US flag, *25*, 92, *143*–146, 161
USS *Hornet*, 167–168
US State Department, 174
US Supreme Court, 130

V-2 rocket, *8*–9, 10–11, 28, 65
Vanguard rocket (US), 11–13
Venus, mission planning, 14
Verne, Jules, 27, 92
"Vomit comet," *54*, *57*
Von Braun, Wernher, 9–11, 13, 15, 28, 31, 36–37, 65, 76, 95, 99, 175
Vostok 1 (Soviet rocket), 14
Vought Aerospace, 35

Water, presence on the Moon, 179
Webb, James, 15, 29, 32, *35*
White, Edward H., II ("Ed"), 22, *38*, 41
Wiesner, Jerome, 15
World War II, 9–10, 17, 28, 83

X-15 (rocket plane), 20, *21*, 61

Young, John W., *51*–53

Zero-g training, 54–55, 57, 59

An artist's impression of Armstrong's first look at the lunar surface. While he had his gold-coated visor down for much of the Moonwalk, he did lift it to look into shadows and dark areas. This image, based on Armstrong's later description, captures what those first few moments might have felt like.